5th International Conference on Upstream Engineering and Flow Assurance 2017

Topical Conference at the 2017 AIChE Spring Meeting and
13th Global Congress on Process Safety

San Antonio, Texas, USA
26 – 30 March 2017

ISBN: 978-1-5108-4133-8

Printed from e-media with permission by:

Curran Associates, Inc.
57 Morehouse Lane
Red Hook, NY 12571

Some format issues inherent in the e-media version may also appear in this print version.

Printed by Curran Associates, Inc. (2017)

For permission requests, please contact AIChE
at the address below.

AIChE
120 Wall Street, FL 23
New York, NY 10005-4020

Phone: (800) 242-4363
Fax: (203) 775-5177

www.aiche.org

Additional copies of this publication are available from:

Curran Associates, Inc.
57 Morehouse Lane
Red Hook, NY 12571 USA
Phone: 845-758-0400
Fax: 845-758-2633
Email: curran@proceedings.com
Web: www.proceedings.com

TABLE OF CONTENTS

482030 AGILE Award Keynote Address

Abstract:

Factors in Selecting AGILE Honoree

1. Overall sustained excellent industry leadership in businesses that employ chemical engineers

2. Recognized for efforts in innovation and creating strategies that ensure success

3. Value of the keynote to the meeting attendees and the members of the Institute Schedule

Event	Time	Location
Continental Breakfast	7:00AM - 8:00AM	Henry Gonzalez Convention Center, Lila Cockrell Theatre
AGILE Keynote Address	8:00AM - 9:15AM	Henry Gonzalez Convention Center, Lila Cockrell Theatre

The 2017 AIChE Spring Meeting and 13th GCPS AGILE Keynote Address will be presented by AGILE award recipient John Y. Televantos, Partner, Arsenal Capital Partners.

John Televantos is a Partner at Arsenal Capital Partners, a middle market private equity firm focused on the specialty industrial, healthcare and specialty financial service sectors. Before joining Arsenal in 2006, Dr. Televantos was President of Aqualon Company, owned by Hercules, Inc.; President and Founder of Helios Chemical Company; and Chief Executive Officer for Foamex International.

Dr. Televantos began his career with Union Carbide in 1977 in silicones research. There he developed two key processes for the manufacture of Methyl Chloride and Siloxane polymers that were commercialized. These processes not only improved efficiency and reduced cost but also dramatically reduced waste. He moved through a variety of management positions there and later at ARCO Chemical, where he became vice president of strategy and development.

At ARCO John co-invented and led a team that developed and commercialized a breakthrough technology in polyol manufacture (a key raw material for polyurethanes) eliminating a whole process step along with waste and emissions, while at the same time opening the door to the creation of new higher-performance polyurethanes. The technology has since been licensed by most of the major global polyol manufacturers with significant waste reduction worldwide.

At Arsenal Capital, Dr. Televantos serves as co-head of the specialty industrial practice. During his tenure, he has been part of several acquisitions and successful exits in the chemicals and materials sectors. In addition to serving on the boards of several Arsenal-owned companies, he is lead director on the board of Balchem Corp.

Unconventional Resources Session in the Upstream Engineering and Flow Assurance (UEFA) Forum

Fluid Report Quality Control and Characterization for the Avalon Shale in the Permian Basin

481 477
Monday, March 27th 2017

Sandhya Sundar-Ram, Ph.D
Senior Reservoir Engineer

Contributions from: David S Miller, Brice Peterson & Birol Dindoruk

Shell Exploration and Production Company

Outline

Brief background of the Avalon play within the Permian Basin

Fluid sample analysis and key findings

Summary of key business impacts

Permian Basin – Avalon play

Grounding Statements

Fluid Properties from recombination samples

Areal Variation of fluid properties

GOR trend of producing wells

3

Permian – Delaware Basin – Regional Setting

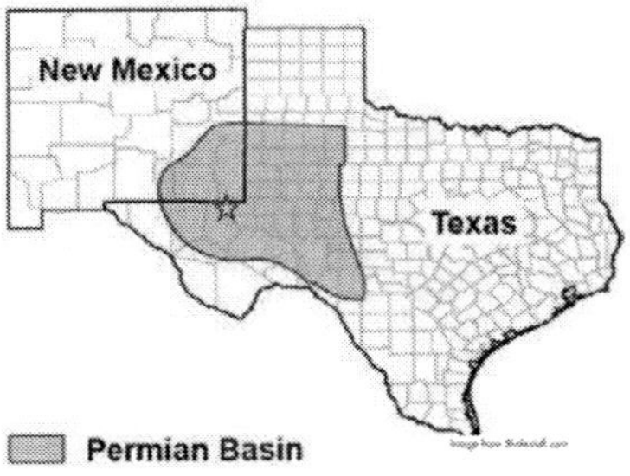

- The Permian Basin is situated in the west part of the state of Texas and the South east New Mexico.
- The main cities where operations are run out of the cities of Midland and Odessa.
- It is a sedimentary basin with one of the world's thickest deposits from the Permian time frame.
- The Permian Basin has produced over 29 billion barrels of oil and 75 trillion cubic feet of gas over several decades.
- Currently, it has more than 3000 O&G operators and has seen ~ $ 50 billion in M&A activity in the last few years
- This presentation looks at data from the Delaware basin mostly around Loving County.

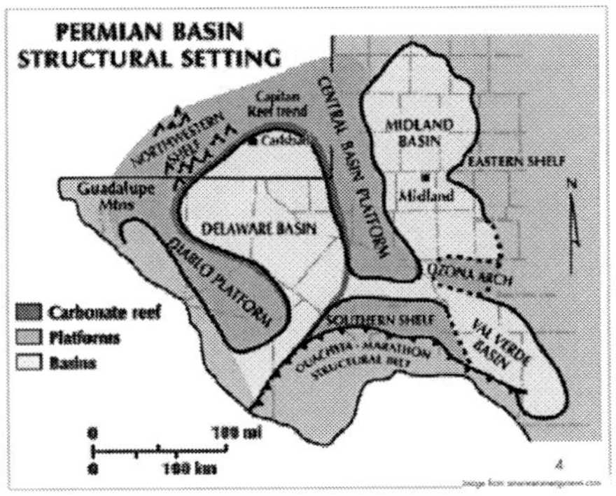

4

Permian Basin Type Log

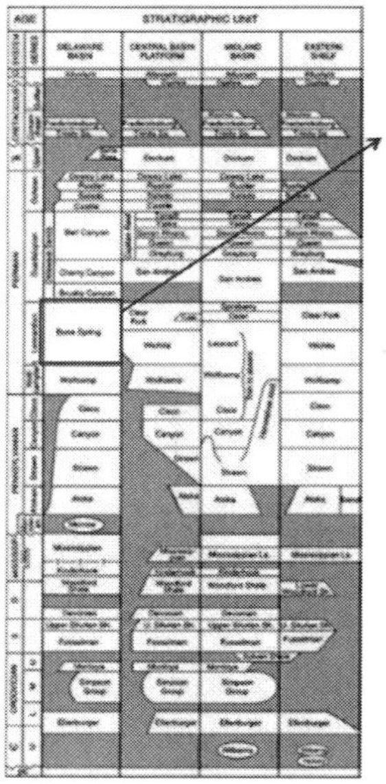

Upper Avalon Shale

- Avalon or Avalon Shale, an informal industry name, refers to upper most part of Bone Spring (BSPG) Formation
- Avalon represents waning phase of BSPG turbidite depositions; sands deposited during low stands, carbonates and shale are high stand deposits
- ~200 – 450 ft thick
- Bounding Carbonates sometimes frac barriers / sometimes breached
- Locally contains ~30 ft wet sand

Evaluation of whether the play was gas condensate or oil

High GOR area described by gas type curve and associated CGR

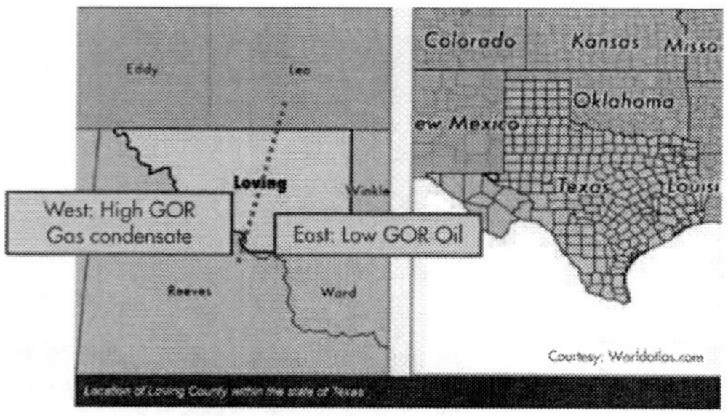

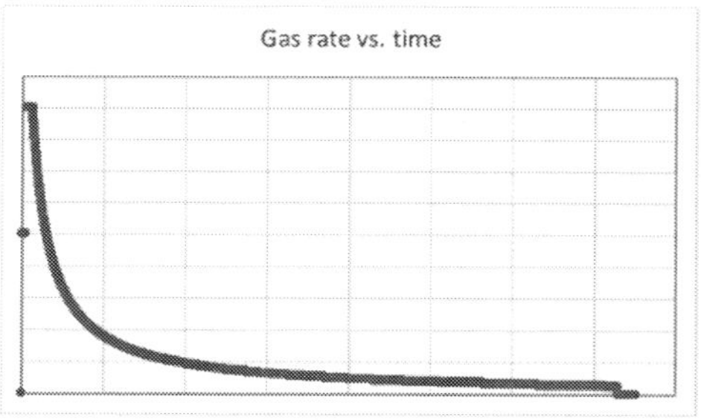

- Regional trending of GOR increasing from east to west
- Most industry view was aligned that the east area was a low GOR oil play
- However, no consensus on fluid type in the west area (gas condensate or volatile oil ???)
- Fluid type impacts business decision and operating procedures → decision to re-evaluate upon acquiring the assets

Available Fluid Sample data

Fluid Reports	Well #1 (east)	Well #2 (west)	Well #3 (west)
Reservoir pressure - psia	hydrostatic	hydrostatic	hydrostatic
Reservoir (sample) temperature – °F	118	136	132
Saturation point - psia	Pres > Psat	Pres < Psat	Pres < Psat
Solution GOR – scf/stb	1,573		3,375
Stock tank gravity - API	42.1		47.4

Fluid samples and production data indicated two kinds of fluid present

Low GOR to the East, High GOR area to the west of the play described by gas type curve

Observations from reported data

2 out of 3 samples indicate Pres < Psat

is this real?

if no – If the reservoir is at or near bubble point (or dew point)

– a fluid sample taken will always be compromised, (finite drawdown)

This drawdown is needed to propel the oil and gas molecules to the sample point.

- so at the sample point one will always be below bubble or dew point.

– In the case of an oil reservoir, sampling below bubble point tends to give too much gas in the sample because gas is much more mobile than oil.

If there is too much gas then we recombine at a too high GOR and consequently the bubble point will be higher than the initial reservoir pressure.

Probably the reason why the samples #2 & #3 reported a Psat > Pres.

Can we try and recreate the fluid phase envelope reconstructed from compositional data to verify ?

Looking to production trends to help

Fluid sample analysis

2 of 3 samples indicated Pres < Psat? Is this real?

Was recombination GOR accurate?

Alternate methods to recreate reservoir fluid

Confirmation with samples obtained later

Sample date vs. Well history

are the GOR used for recombination representative of the fluid?

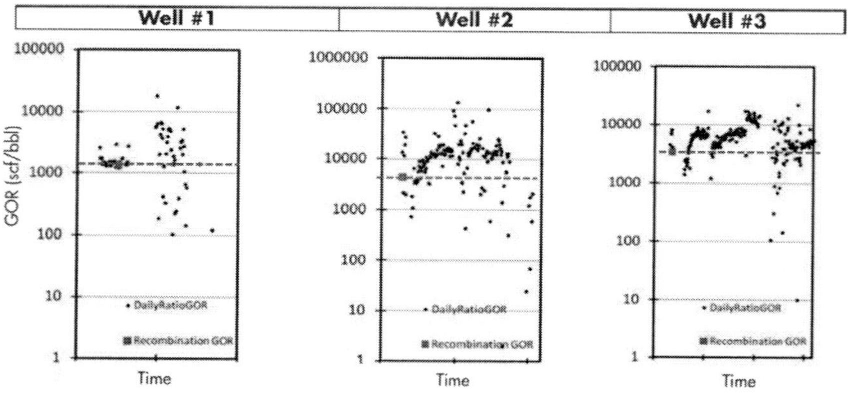

GOR used for recombination: 1365 GOR used for recombination: 4400 GOR used for recombination: 3375

Sample date vs. Well history

are the GOR used for recombination representative of the fluid?

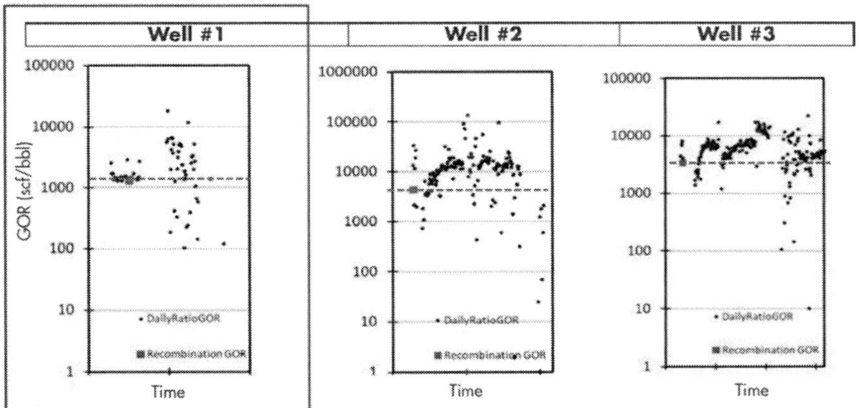

GOR used for recombination: 1365 | GOR used for recombination: 4400 | GOR used for recombination: 3375

Well #1

- Quality check on fluid sample data was carried out
- Sample tuning using software package
- Generation of the phase envelope for the tuned fluid

CCE Saturation Point Check

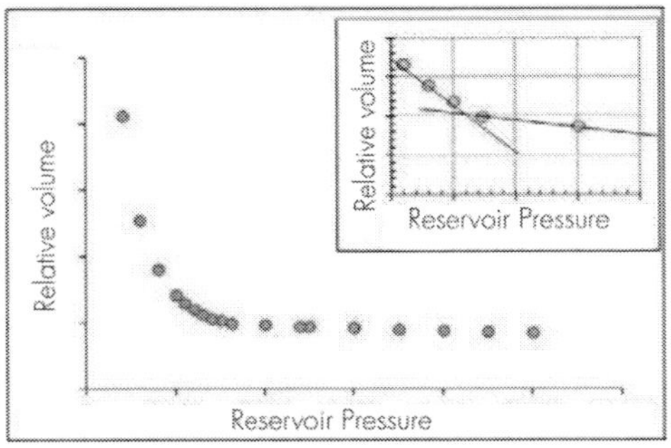

- ✓ Intersection point fuzzy; typical for a volatile oil
- ✓ Contractor took sufficient data points
- ✓ Reported bubble point that was consistent

Phase envelope of tuned fluid

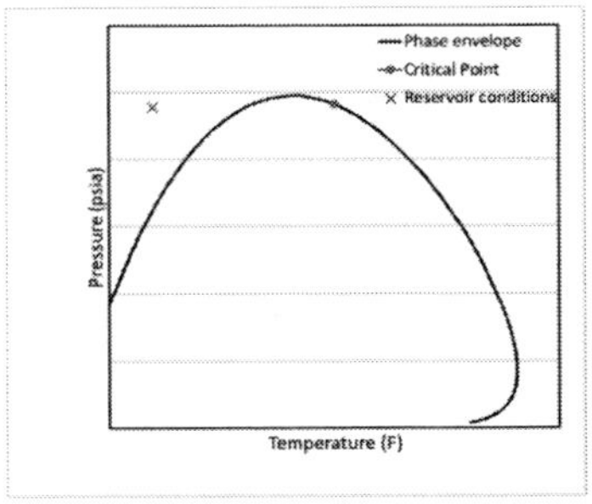

Key Learning: Avalon play is in the east is a confirmed to be an oil play consistent with sampling and production data

Sample date vs. Well history

are the GOR used for recombination representative of the fluid?

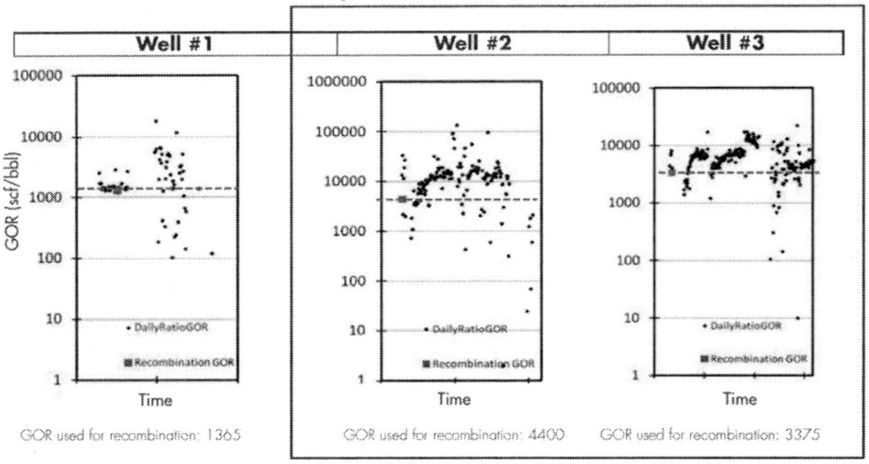

Attempts to reconcile data for Well #2 & Well #3 to the west

With the sampling pressure being below bubble point its likely that an excessive amount of reservoir gas has been sampled.

We can make an attempt to correct as follows:

- Review any other source of reservoir pressure or regional gradients and obtain a representative reservoir fluid gradient in the location of the sampled well.
- Use the separator gas and liquid samples and recombine at different GORs to recreate the reservoir fluid
- Two ways forward:

1. Either match on reservoir pressure (If we believe that the fluid is indeed at saturation conditions initially)
2. Or use the most representative solution GOR for the fluid at recreate reservoir fluid

Observations and Recommendations

Production data indicates that recombination GOR may be too high.

Testing range on uncertainty:

1. Input fluid compositions in PVTSim®
2. Recombine at stock tank (residual) oil density for a range of GOR to recreate fluid at reservoir conditions close to bubble point
3. Plot the phase envelopes with critical point and include reservoir pressure

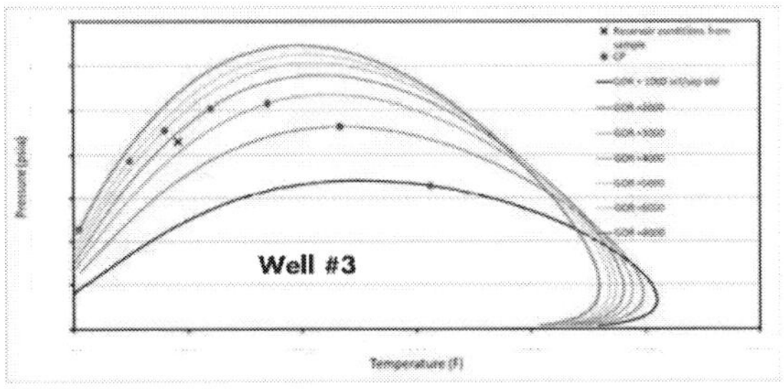

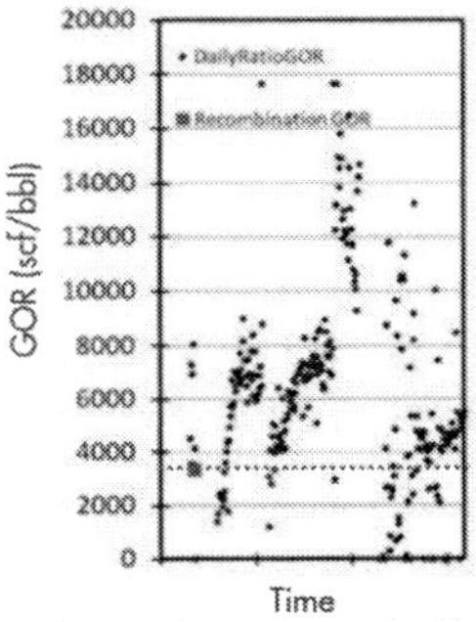

Key Learning: Avalon play is most likely a Volatile Oil play with Bubble point close to Pres with a lower GOR in the west

Next Sample planned was recombined more carefully

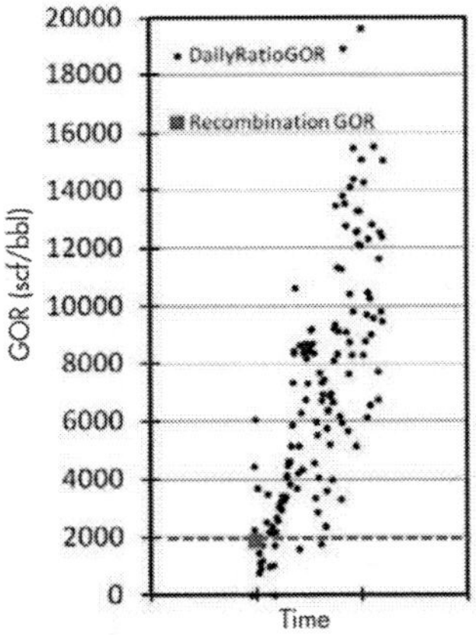

Fluid Reports	Well #4
Reservoir pressure - psia	hydrostatic
Reservoir (sample) temperature – °F	125
Saturation point - psia	Pres > Psat
Solution GOR – scf/stb	1867

Confirmation: Avalon play is a Volatile Oil play with a regional variation in GOR

Well #4

- Quality check on fluid sample data was carried out
- Sample tuning using software package
- Generation of the phase envelope for the tuned fluid

CCE Saturation Point Check

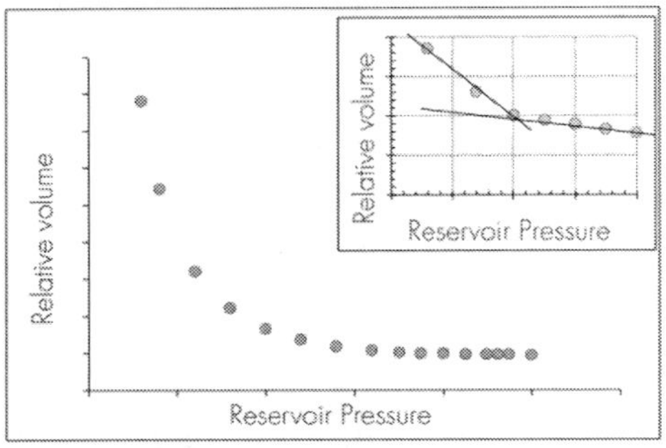

✓ Intersection point fuzzy; typical for a volatile oil

✓ Contractor took sufficient data points

✓ Reported bubble point consistent with data

Phase envelope of tuned fluid

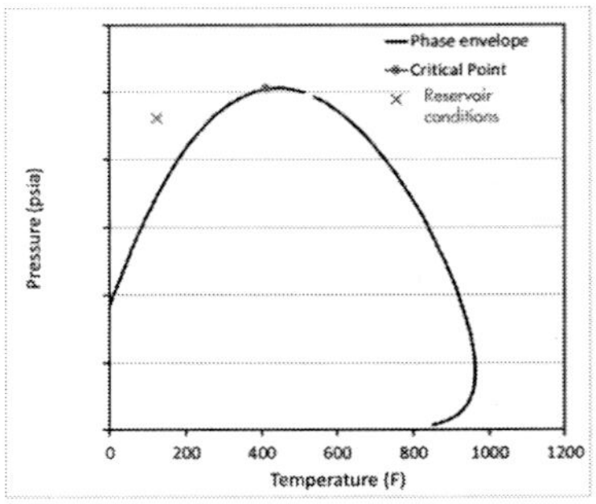

Summary and Key business impact

Avalon play is a Volatile Oil play with a regional variation in GOR increasing from east to west

Business impact due to improved economics of oil play vs. gas condensate play

Updated choke management to maximize recovery from reservoir

 18

Prior View

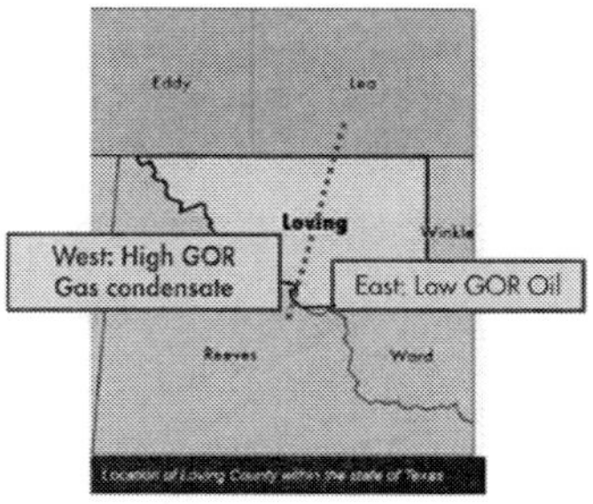

Updated View

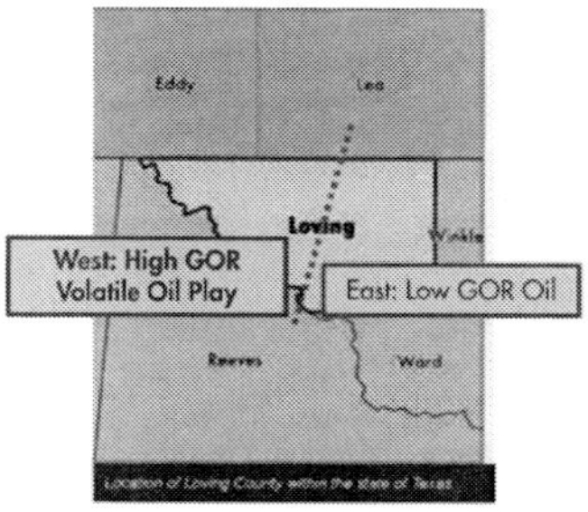

Updated PVT analysis – west area was a volatile oil play with bubble point close to initial reservoir pressure

New type curve was generated – oil with associated GOR → Positive impact to resources and capital justification for planned wells

Key Impact to operations: updated choke strategy to maximize recovery and minimize liquids left behind

Back Up Slides

Consequences of condensate drop-out

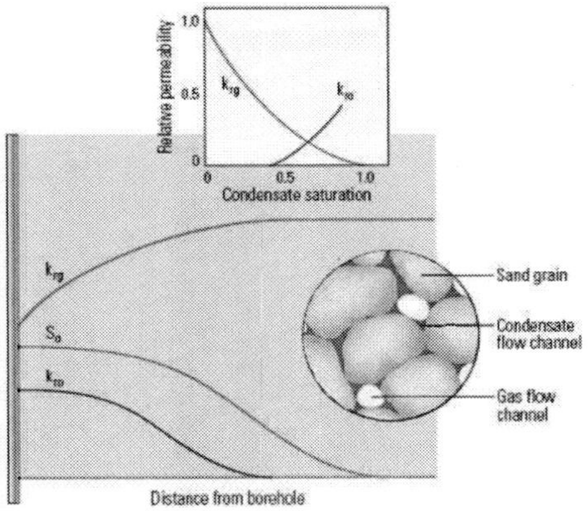

Even for rich gas condensates with substantial drop-out, the condensate mobility remains insignificant away from the well-bore.

- The condensate that forms in most of the reservoir is lost to production unless the depletion plan includes cycling.

Near a producing well the situation is different: enough liquid accumulates that its mobility becomes significant, but less flow path is available for gas.

- Condensate blockage is a result of the decreased gas mobility around a producing well below the dew-point.

482168 Laboratory Evaluation of a Natural Gas-Based Foamed Fracturing Fluid

Abstract:

Foams have been used as hydraulic fracturing fluids to reduce water usage and minimize the potentially deleterious impact on water-sensitive formations. Traditionally, carbon dioxide (CO_2) and nitrogen (N_2) have been used as the internal phase in these foamed fluids. Hydraulic fracturing with natural gas is a relatively inexpensive option, particularly if natural gas produced from the wellhead can be used without significant processing. As part of a Department of Energy (DOE) sponsored program, we have previously identified an optimal thermodynamic pathway to transform wellhead natural gas into pressurized natural gas suitable for use as the internal phase in a foamed fracturing fluid.

The current study is focused on preparing a natural gas-based foam at surface conditions typically encountered in hydraulic fracturing and measuring the stability and rheological properties of the foam. In addition, the transient response of the foam and the pumping equipment used to break the formation down was simulated using a burst disc. A base fluid was prepared by batch mixing a viscosifier and foaming surfactant with water. The base fluid was then injected into a tee using a water pump. Simultaneously, liquefied natural gas (LNG) was pressurized using a cryogenic pump, vaporized using a heat exchanger, and injected into the tee to contact the base fluid and generate a foam. The mixture was then pumped through approximately 300 ft of 0.312-in ID tubing equipped with pressure transducers at several locations. The test fixture included a sight glass to visually determine the quality of the foam. Fluids were pumped at various natural gas/base fluid ratios (60 to 80% foam quality), pressures (2,500 to 7,500 psi), and flow rates (0.5 to 7 galUS/min); the resulting pressure gradients in the tubing were measured at each test condition. This presentation reports our findings related to foam stability and rheology and compares these results to previous studies on foamed fracturing fluids.

479199 Self-Repairing Cement Composites for Subsurface Applications

Abstract:

Wellbore failure at the cement lining is one of the most common drivers of reservoir intervention during geothermal energy and unconventional oil/gas production. The main causes for cement fracturing are chemically corrosive (typically hypersaline, CO_2 and H_2S-rich) environments, mechanical stress, and high temperatures (up to 190 °C in unconventional oil reservoirs and up to 350 °C in Enhanced Geothermal Systems). As a result, expensive and time-intensive production shutdowns and repairs are required. Intervention costs average $1.5million per wellbore without taking into consideration the economical losses as a result of production stoppage, which can be several million dollars depending on the time frame of plant in non-production mode. To address these problems we developed a thermally stable polymer-cement composite with self-healing properties while maintaining the required rheological (during pumping) and mechanical properties of typical wellbore cement. Characterization of these cementitious materials comprised evaluation of rheological properties, including density and consistency measurements; mechanical properties, including compressive strength, Young modulus, and fracture toughness; thermal, chemical, and morphological properties, SEM/EDX, X-ray diffraction, FTIR, total organic carbon, and TG-MS. We demonstrated that these novel composite formulations present all the requirements of standard wellbore cement and introduce self-healing capabilities as shown by reducing by up to 82% the permeability of mechanically-induced fractures in the 0.3-0.5mm aperture range. These polymer-cement composite materials could then represent a definite solution to wellbore failure, production stoppage and reservoir intervention during geothermal and fossil energy production. In this presentation we will briefly describe the concept and present experimental as well as modeling results obtained this far on this latest technology.

480206 Effect of Low-Dosage Aminoacids Based Kinetic Hydrate Inhibitors

Abstract:

Gas hydrates are metastable crystalline compounds that are formed when small gas molecules like ethane and methane gets trapped within the water molecules, at higher pressure and lower temperature conditions. These gas hydrates tend to accumulate in the offshore subsea lines and over the period tend to completely block the pipeline, halting the gas and oil supply to the onshore stations. Therefore, annually the industry spends over 100 million dollars on hydrate prevention procedures. These procedures include mainly the use of chemical thermodynamic hydrate inhibitors (THI) and Kinetic hydrate inhibitors (KHI). The THI like Mono-ethylene glycol (MEG) and Methanol are mainly used in industry. However, these THI are required in bulk quantities, which causes a major disposal issue and also post a major safety threat to nearby operators due to high flammability of Methanol. These commercial THI are also not suitable for environmental, so there is a strong research interest in finding a new class of low dosage KHI and THI inhibitors that are environmentally friendly and effective as commercial THI and KHI.

Amino acids are biological organic compounds containing amine ($-NH_2$) and carboxylic acid functional groups (-COOH) that have been previously reported to be an environmentally friendly corrosion inhibitors. So the feasibility of using amino acids as a potential KHI and THI have been tested using pure methane gas on a rocking cell assembly within the pressure range of 40-120 bars. The types of amino acids tested include: L-Alanine, L-Phenylalanine, Glycine, Histadine and L-Asparagine.

The experimental results obtained show that the amino acids with lower solubility provided significant KHI effect, but no THI effect. However, the amino acids with higher solubility provided both THI and KHI effect simultaneously. The amino acids with higher solubility were able to shift hydrate vapour liquid equilibrium curve (HLVE) by 0.7 ^{o} C at both lower (~40 bars) and higher (~100 bars) pressures. These amino acids were also able to delay methane hydrate crystal formation by up to 1 hour at both lower (~40 bars) and higher (~100 bars) pressures. However, the addition of synergents with amino acids delayed the hydrate inhibition by almost 24 hrs at lower (~40 bars) and by 6 hrs at higher (~100 bars) pressures. Thus, the selected amino acids with higher solubility were found to have a have a potential to act as a KHI and THI both simultaneously.

This work was made possible by NPRP grant # 6-330-2-140 and GSRA # 2-1-0603-14012 from the Qatar National Research Fund (a member of Qatar Foundation). The statements made herein are solely the responsibility of the authors.

480409 Development and Experimental Validation of Hydrate-Philic Surfactants

Abstract:

Clathrate hydrates are crystalline inclusion compounds which can form in subsea flowlines and lead to the formation of blockages. To mitigate the risk of hydrate blockages, conventional flow assurance approaches have deployed thermodynamic hydrate inhibitors (THI), such as methanol or monoethylene glycol, to move the pipeline out of the hydrate stability region. In deepwater operations, there is an increased driving force for hydrate formation and the amount of THI required to prevent hydrate formation in the flowline may be cost prohibitive. To maintain the viability of deepwater assets, alternative hydrate management strategies are required; anti-agglomerants (AAs) are one promising field, where surfactants are used to prevent hydrate particles from aggregating in the liquid phase or depositing on the flowline wall. However, the adoption and deployment of AAs may be limited, as the ionic nature of current-generation AAs may generate strong water-in-oil emulsions and may also have a deleterious effect on water quality. The development of surfactants that selectively adsorb to the hydrate crystal surface may remedy both operating challenges, but requires an experimental technique to characterize the hydrate-philicity of these chemicals. We present a new experimental protocol using a micromechanical force (MMF) apparatus and a pendant drop interfacial tensiometer to respectively measure hydrate cohesive force and oil-water interfacial tension. We have deployed this protocol to characterize several non-ionic surfactants, thereby identifying chemicals that reduce the hydrate cohesive force to a required level with minimal penalty to the water-oil interfacial tension. The application of this experimental protocol, based on an initial set of experiments, provides an attractive pathway to tailor the next generation of hydrate management chemicals, which are of particular importance in sensitive marine environments.

ESTIMATION OF METHANOL PILL SIZE FOR HYDRATE PREVENTION DURING OFFSHORE OIL & GAS SUBSEA SYSTEM OPERATIONS

Ramesh A. Kini, Chevron Energy Technology Company, Houston, TX; Juan J. Manzano-Ruiz, Bechtel OG&C, Houston, TX; and Steven W. Cochran, Chevron Energy Technology Company, Houston, TX.

Introduction

Subsea oil production from offshore reservoirs faces substantial risk of hydrate blockage formation while carrying out routine operations such as dead-oil displacement (DOD) from a dead-leg line into a producing line, or a restart of the subsea system after an unplanned shutdown. The operating company usually forgoes continuous injection of hydrate inhibitors (e.g. methanol, mono-ethylene glycol, LDHI) to avoid fluid contamination and minimize operating expenses. Instead of continuous inhibitor injection, oil production streams are typically insulated for normal operation; however, that leads to more complicated operations for shutdowns and restarts.

The common practice among deepwater oil producers is to run uninhibited for normal operation and, in case of an unplanned shutdown, a dead-oil displacement procedure is executed to displace live oil within the subsea system with dead oil. The dehydrated and gas-free dead oil can withstand a long shutdown at any subsea pressure/temperature without hydrate formation. Once the shutdown is over, the trapped dead oil is delivered to the topsides and the production of live oil can be safely resumed. The DOD operation has to be executed soon after shutdown, typically within 8 to 12 hours after the flow stops, depending on how thermally insulated the flowlines/risers are. During the DOD operation there is a risk of hydrate formation at the interface between the dead oil and cold live oil due to the increase in pressure caused by the pumping needed for displacement. To prevent a potential hydrate issue at this interface a batch of methanol (i.e. a "pill") is placed, typically at the subsea manifold (Figure 1), ahead of the dead-oil to separate the two streams. In subsea systems with long flowlines and deep risers, the methanol pill is usually oversized to be conservative [1]. Such a large volume of methanol is a concern for the operator not only for its cost, but because it depletes the limited inventory onboard the topsides and contaminates the resulting export gas and oil streams leading to out-of-spec products.

This paper focuses on establishing an alternative method to size the methanol pill to avoid oversizing, as typically observed in oil and gas operations. The optimum size depends on the mechanism of methanol phase dispersion by turbulent diffusion through the phases fore and aft the pill, methanol loss through thin-film shedding behind the travelling pill followed by the pushing phase, and methanol retention due to gravity override. This paper evaluates the turbulent mass-transfer and hydraulics models to estimate the volume of methanol lost to the adjacent fluids during the full transit of the pill from injection point to the topsides. The basis of mass-transfer analysis is the momentum- and mass-transfer Reynolds-analogy to derive a virtual turbulent-coefficient of diffusion "K", following G.I.Taylor's 1954 seminal paper [2].

A stability analysis of the potential stratification of the methanol phase overriding the oil phases was carried out based on Richardson-number estimation. For the typical subsea applications herein addressed, the Richardson-number was less than 0.8, indicating that stratification is very weak or unlikely, whilst turbulent diffusion governs the flow process. For high Reynolds-number flows, turbulence stresses will dominate and all stratification concerns vanish. Based on these arguments the potential flow pattern distortion caused by phase stratification was neglected in this study.

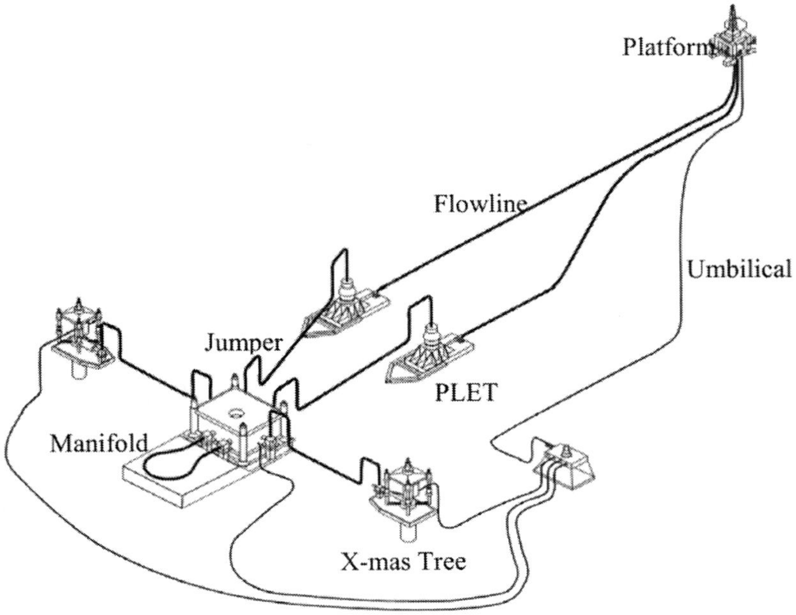

Figure 1: A Schematic of Subsea Layout

Examples of a typical DOD operation with a methanol-pill between dead oil and live oil were analyzed in this work. The mixing at the methanol pill boundaries with live oil and dead oil was modelled with the transient diffusion PDE and the diffusion K-coefficient for 4- to 5-mile long flowlines with diameters ranging between 6 and 10 inches. The resulting error-function concentration distribution for the methanol/live oil interface yielded mixing layers ranging roughly between 160 and 640 feet long, depending on the prevailing Reynolds number. Similar length-scale results were also obtained for the methanol/dead oil interface. The mixed volume of methanol with dead- and live oil varies between 3 and 42 bbls for these examples. The methanol film-thickness in these examples was estimated assuming that the Reynolds analogy guarantees that the fluids dispersion in the radial direction is very effective to the level where the turbulent-diffusion balances out with the molecular-diffusion, i.e. within the "buffer" layer of the turbulent velocity profile. With the latter assumption an estimate for film thickness was determined as a function of Reynolds number and shear stress. Typical results obtained for the film volume are of the order of 4 – 8 bbls.

Flow Models

The interface between the methanol-pill and the driven fluid will be deformed by the velocity profile developed. In the case of turbulent flow, the velocity profile is almost flat over the majority of the cross section. A logarithmic velocity distribution close to the pipe wall represents well the approach to zero velocity at the wall itself.

Figure 2 shows a sketch of the expected velocity profiles for the methanol interfaces with dead oil and live oil (driven fluid). The velocity profile is superimposed over the mixing zone where methanol and oil blend by turbulent dispersion.

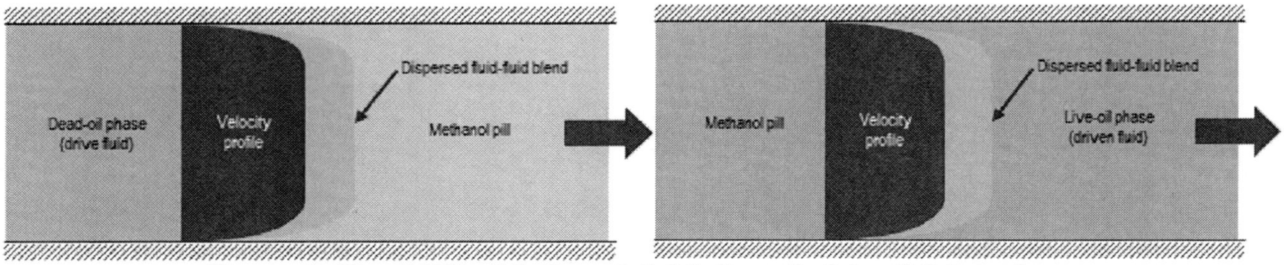

Figure 2: Dead Oil – Methanol & Methanol - Live Oil interfaces

Within the mixing zone the phases' concentration distributes following an error-curve shape, characteristic of diffusional processes. In this case the dispersion of soluble fluid is driven by a virtual coefficient of turbulent diffusion to be later defined.

In the DOD operation the dead-oil phase will push the pill to displace the live oil minimizing the hydrate formation risk due to cold dead oil contacting the live fluid at increased pressure. A similar mixing zone will form at the back of the methanol pill with the dead-oil phase as depicted in Figure 2 as well.

These flow models at both ends of the pill are useful to visualize the shrinkage of methanol pill that ends blending with both the dead and live oils. Such blending processes contribute to shrink the initial batch volume of methanol, which becomes a contaminant of the oil phases. The blending volumes are accounted in this methodology.

Turbulence Diffusion Model

This section reviews the original Taylor's model [2] which is the basis for this study. Some typographical corrections were made to the original publication and an improved numerical-integration scheme was carried out. Taylor's paper included a demonstration that the dispersion of diffusible matter in turbulent flow, from a source travelling at time-average speed, can be modelled by a virtual coefficient of diffusion (i.e. "K"). The demonstration is based on the analogy between the turbulent dispersion correlation and the mass-diffusion equation:

$$\frac{\partial C}{\partial t} = K \frac{\partial^2 C}{\partial x_1^2} \qquad (1)$$

where C is the concentration per unit volume and x_1 is the longitudinal coordinate measured on a reference frame travelling at the average speed $\bar{U}$. From this analogy Taylor concluded that the coefficient of diffusion correlates with the time-average of the square of the axial velocity fluctuations. Following Taylor's model, we reviewed the estimate of the K coefficient to assess the time evolution of the phases' dispersion, i.e. the growth of the mixing zones fore and aft the methanol pill.

To support the turbulence model, Taylor [2] used the experimental data for the velocity distribution across the pipe section taken from Nikuradse's 1932 publication [3]. The integration of the resulting diffusion PDE leads, after some corrections in the numerical integration process, to the following expression:

$$K = 10.424au^* \qquad (2)$$

where "a" is the inner radius of the pipe and "u*" is the turbulent shear velocity. The coefficient in Equation (2) was corrected for longitudinal turbulent diffusion (by 0.6%) to finally yield:

$$K = 10.482au^* \approx 10.5au^* \qquad (3)$$

Model Validation

Previous experiments performed by Allen and Taylor in 1923 [4] for salt dispersion in water solution allowed to infer a turbulent coefficient of diffusion of 10.6, very close to the above result of Equation (3). Further experiments performed at the Cavendish Laboratory [2] through rough piping yielded a turbulent coefficient of diffusion of 10.5 matching Equation (3) and demonstrating the applicability of this methodology to both smooth and rough wall piping.

Methanol-Film Thickness Estimate

The methanol pill will leave a thin wetting-film adhered to the inner pipe wall during the dead-oil displacement operation. This wetting-film represents the loss of methanol to inhibit the live-oil phase moving ahead. A model is proposed to estimate the volume of methanol film from the pill based on the typical turbulent flow velocity profile. The model focuses on the ratio of the "defect" flowrate associated to the film and the net fluid flowrate through the flowline.

Figure 3 illustrates three major fluid volumes, namely:

- The volume of methanol pill pushed by dead oil, represented by the "orange" area,
- The film of methanol trapped underneath the penetrating flow of dead oil, forming the trailing wetting film and represented by the "green" area,
- The volume of methanol pushed forward, close to the pipe wall, herein represented by the "blue" area and subject to axial turbulent diffusion.

By continuity the net fluid flowrate through the flowline is given by:

$$Q_{net} = \int_0^{R_o} 2\pi r u(r) dr = \pi R_o^2 \overline{U}_{avg} \qquad (4)$$

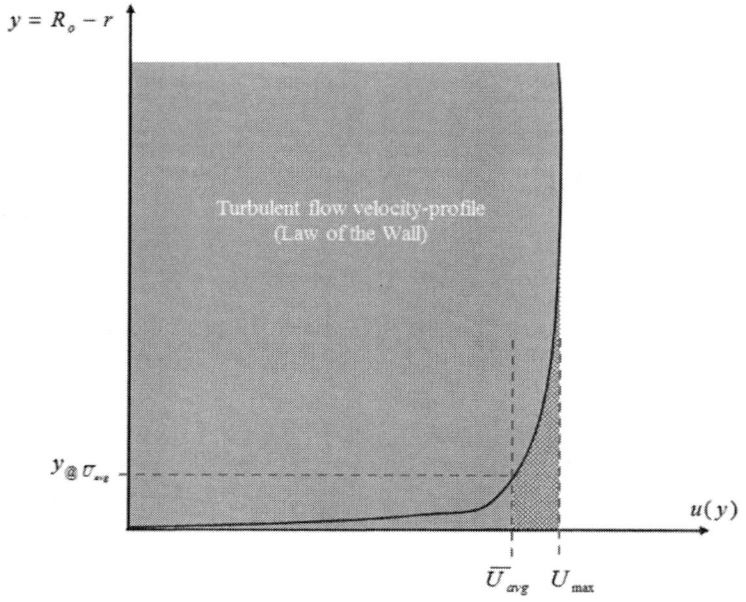

Figure 3: Core flow (orange area) and film defect-flow areas (green area)

To carry out the integral in Equation (4) we used Martinelli's [5] explicit correlations for the three distinctive turbulent-layers extending from the wall to the center-line:

- $u^+ = y^+$ for $y^+ \leq 5$ (laminar sublayer) (5)
- $u^+ = 5Ln(y^+) - 3.05$ for $5 \leq y^+ \leq 30$ (buffer layer) (6)
- $u^+ = 2.5Ln(y^+) + 5.5$ for $y^+ \geq 30$ (turbulent core to centerline) (7)

The film flowrate (green area of Figure 3) is equivalent to the flowrate associated to the turbulent velocity profile bound between the maximum and the average velocities; this flowrate is given by:

$$Q_{film} = \int_{r_{U_{avg}}}^{R_o} 2\pi r u(r) dr \qquad (8)$$

The methanol-film volume can now be estimated from the flowrate ratio of Equations (4) and (8) and the original pill size as:

$$V_{film\ MeOH} \approx \left(\frac{Q_{film}}{Q_{net}}\right) V_{pill} \qquad (9)$$

This film-volume estimate has been calculated for a range of typical dead-oil displacement operations and correlated to the Reynolds number. The ratio of the flowrates (Q_{film}/Q_{net}) is presented in Figure 4. Depending on the Reynolds number of the flow, the methanol in the wetting film ranges between 4.5 – 7.5% the original pill size. Experimental validation of the loss of methanol due to film formation is planned in a future phase of work.

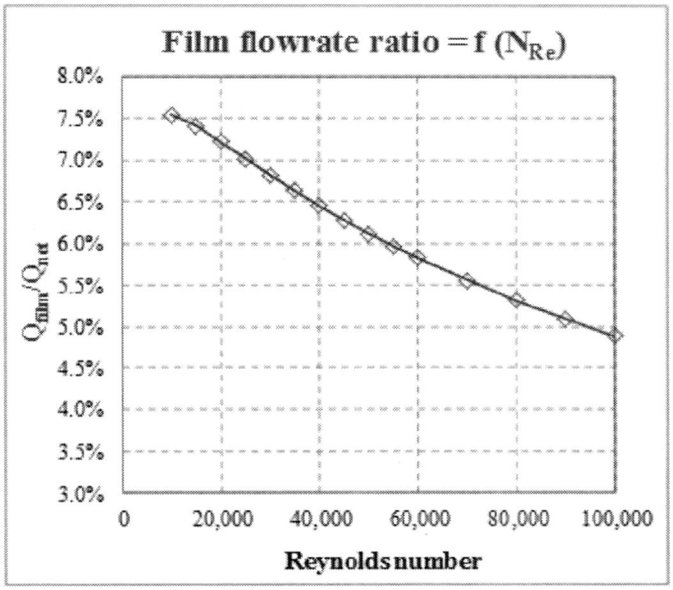

Figure 4: Ratio of film defect-flowrate to net flowrate

Fluid Concentration at Methanol-Live Oil Interface

Methanol will blend with live oil at the front end of the pill spreading the beneficial inhibiting effect on the oil phase. This blending will be enhanced by the turbulent dispersion of the phases. As a result of such blending the size of the pill correspondingly will shrink in volume.

The phases' concentration profiles can be estimated using the error function, erf(z), typical solution for diffusional processes. The concentration of matter is calculated by:

$$C_{MeOH} = \frac{1}{2}\left[1 - erf\left(\frac{x_1}{\sqrt{4Kt}}\right)\right] \qquad (10)$$

where "K" is the virtual coefficient of diffusion (matching the corresponding phase); "x_1" is the distance measured from the reference frame moving at the average speed; and "t" is the time from the operation start-up. A typical blending profile is given in **Figure 5**.

To estimate the length of the blend region requires a somewhat arbitrary definition. Following Taylor, the length of blend region is defined as where the methanol concentration changes from 99% to 1% at the interface. With this definition the length "$S_{MeOH-LO}$" of methanol-oil is approximately:

$$S_{MeOH-LO} \cong 1.645\sqrt{4t}\left(\sqrt{K_{MeOH}} + \sqrt{K_{LiveOil}}\right) \qquad (11)$$

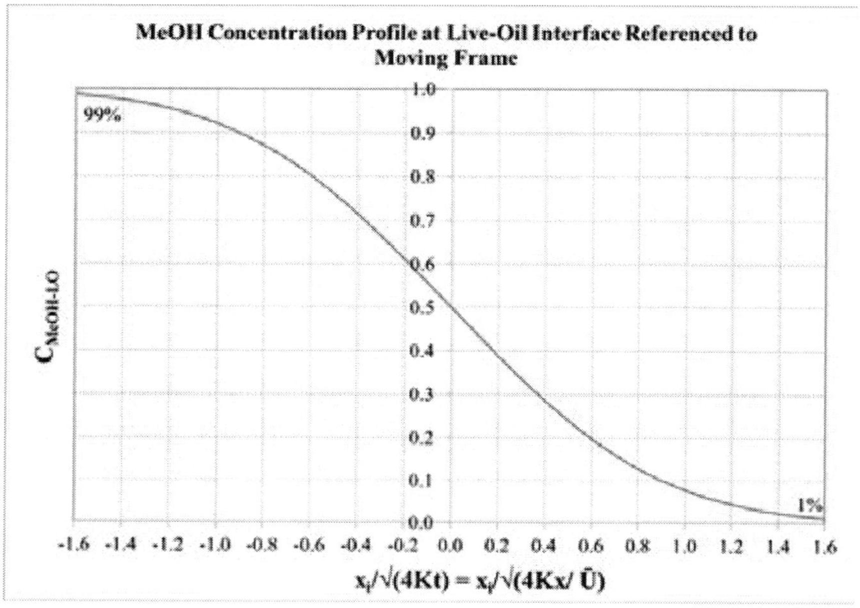

Figure 5: Blending profile at live oil & methanol interface

The corresponding volume of blended methanol with live oil will be:

$$V_{MeOH,front} = 0.5 \int_{x_{1,99\%}}^{x_{1,1\%}} \left[1 - erf\left(\frac{x_1}{\sqrt{4Kt}}\right)\right] dx_1 \qquad (12)$$

The models presented in this work apply for established steady-state flow only. The live-oil phase is assumed to be pseudo-homogenous, and hence methanol partitioning in the live oil phase is not accounted for in our models.

Fluid Concentration at Dead Oil - Methanol Interface

Similarly, at the dead-oil interface, the rear end of the methanol-pill will experience a loss. This process is illustrated in Figure 6, based on the following error function distribution:

$$C_{MeOH} = \frac{1}{2}\left[1 + erf\left(\frac{x_1}{\sqrt{4Kt}}\right)\right] \qquad (13)$$

The length of dead oil and methanol blend will extend according to the following relationship:

$$S_{DO-MeOH} \cong 1.645\sqrt{4t}\left(\sqrt{K_{DeadOil}} + \sqrt{K_{MeOH}}\right) \qquad (14)$$

The volume of blended methanol with dead oil will be:

$$V_{MeOH,rear} = 0.5 \int_{x_{1,1\%}}^{x_{1,99\%}} \left[1 + erf\left(\frac{x_1}{\sqrt{4Kt}}\right)\right] dx_1 \qquad (15)$$

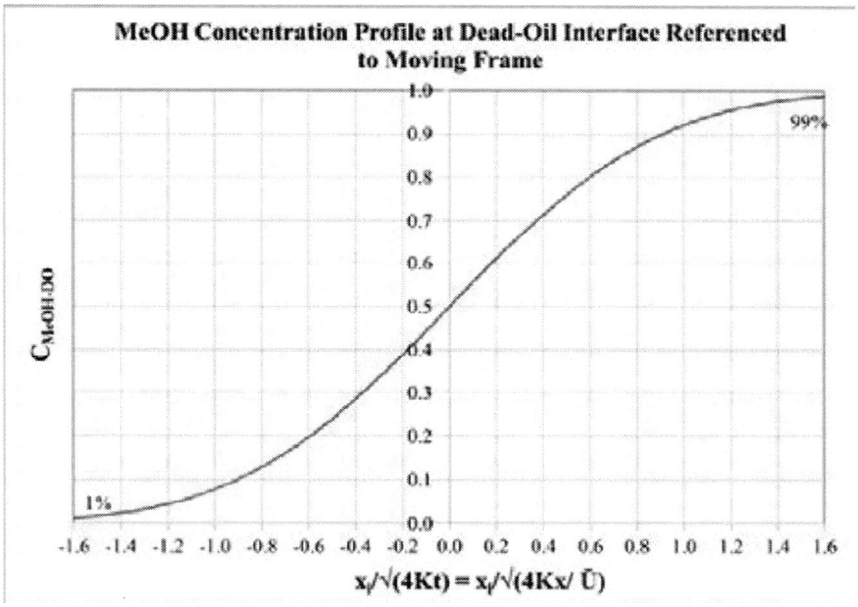

Figure 6: Blending profile at methanol & dead-oil interface

Field Application Example

The following example based on typical dead-oil operations is offered to illustrate the calculations necessary to estimate the safety-factor for the methanol-pill size:

- Live oil: 33.1°API and 4.3 cP
- Dead oil: 28.5°API and 7.2 cP
- Subsea flowline/riser diameter: 7.125" ID (8" nominal)
- Subsea system length: 21,655 feet total (including flowlines and risers)
- Pill size: 100 bbl
- Dead oil flowrate: 25,000 bpd

The following results are obtained:

Table 1: Field Example - Summary of Results

	Dead oil	Methanol	Live oil
Average velocity [ft/s]	5.9		
Reynolds number	39,674	499,711	64,574
Coefficient of dispersion "K" [ft^2/s]	0.976	0.816	0.932
Film-shed flowrate ratio	6.5%		
Wetting-film lost volume [bbl]		6.5	
Front of pill penetration length [ft]		161	184
Rear of pill penetration length [ft]	193	161	
MeOH volume blended w/live oil [bbl]		8.1	
MeOH volume blended w/dead oil [bbl]		9.2	
MeOH volume blended and film shed [bbl]		23.8	
Pill-size safety factor	320% = (100/23.8)-1		

Conclusions

1. The volume shrinkage of the methanol pill caused by wall film wetting accounts for 4.5 to 7.5% of the original pill volume.
2. At low flowrates the extended pill transit time increases the extension of the four blending regions between live oil, methanol and dead oil. The methanol film volume left behind is also higher at low flowrates.
3. At high flowrates the enhanced turbulent flow increases the dispersion constant, but the blending of phases is lower due to the shorter trip time during DOD. Additionally, the film methanol losses are substantially lower at high flowrates.
4. During DOD start-up, laminar flow prevails and the blended volumes extend in length causing higher methanol loss due to the wetting film. The usual operational procedure requires soft DOD start-ups before reaching steady-state causing initial low-flow scenarios. This paper does not address such transient scenarios but provides preliminary analysis for the steady-state case.
5. This method provides the minimum volume of methanol pill based on losses due to diffusion and film formation. The contribution of "gravity override" effect due to the density difference between oil and methanol was found to be negligible.

References

1. Little, H.S., Jessup, J.W., Schoppa, W.W., Seay, M.R., Coulon, T.D. "Perdido Startup: Flow Assurance and Subsea Artificial Lift Performance", OTC 21716, 2011
2. Taylor, G.I. "The dispersion of matter in turbulent flow through a pipe", Proceedings of the Royal Society, A, Vol. 223, pp. 446-468, 1954

3. Nikuradse, J. “Laws of turbulent flow in smooth pipe”, NASA TT F-10, 359; translation from VDI Forschungsheft 356, Edition B, Vol. 3, pp. 1-36, 1932
4. Allen, C.M., Taylor, E.A., Transactions of the American Society of Mechanical Engineers, Vol. 44, p. 285, 1923
5. Martinelli, R.C. Transactions of the American Society of Mechanical Engineers, Vol. 69, p. 947, 1947

479571 Gelation and Thermodynamic Analysis of Ternary Systems of Long-Chain n-Alkanes in a Short-Chain n-Alkane Solvent

Abstract:

The formation of wax deposits in offshore oil pipelines has remained an issue for the petroleum industry for several decades. These wax deposits lead to clogged pipelines that result in significant financial costs for lost production and required maintenance. Although a number of remediation methods exist, there is a lack of complete understanding of how the various components of crude oils affect the crystallization and gelation of the crudes. Of significant interest are long chain n-alkanes, which are typically the major component of waxes because of their low solubility and ease in forming an ordered crystal structure. In order to attempt to gain fundamental insight about what is occurring, model fuels have been utilized consisting of a short-chained n-alkane acting as the solvent and long-chained n-alkanes acting as the solutes. Previous work has shown that both polydispersity and cocrystallization can influence the thermodynamic and gelation characteristics of model oil systems where only two solutes were explored. This work explored systems with three solutes, systems where a mixture of polydispersity and cocrystallization would occur.

To study the influence of cocrystallization and polydispersity, systems containg C28, C32 and C36 were mixed with a dodecane solvent. It is known that C36 is capable of cocrystallizing with C32 and C32 is capable of cocrystallizing with C28. However, C36 is incapable of cocrystallizing with C28. Important parameters such as the cloud point, gel point, pour point and heat of crystallization were determined by the use of rheometry, differential scanning calorimetry and visual techniques. The mass ratios of C36 to C32 was fixed at 1:1, 1:2 and 2:1 while the amount of C28 was varied in the analysis. Work showed that these systems obeyed van't Hoff solubility theory. However, the DSC traces clearly showed that the way in which the materials were cocrystallizing was drastically different as the amount of C28 was increased. Adding C28 to C32-C36 solutions caused a decrease in the gel point of the system. This phenomenon can be explained by the fact that C28 acted as a dispersant and a crystal modifier, preventing the C32 and C36 molecules from interacting with one another and forming the necessary volume spanning network to form a gel. This result was also seen in prior work when C36-C28 systems were analyzed. However, the presence of C32 minimized the magnitude of the decrease of the gel point. These effects were minimized for the system containing 2% C36 and 4% C32. One explanation for why the presence of C28 does not depress the gel and pour points for this system is that there is too much C32 and C28 in the solution relative to the amount of C36, which overcomes the dispersant effect of the C28 on the longer n-alkanes. Another explanation is that since there is less C36 than C32, some C28 can cocrystallize with C32, resulting in more C28 crystallizing at higher temperatures and higher gel/pour points.

480988 Application of Aggregation Thermodynamics and Molecular Characterization to Study Asphaltene Precipitation of Bituminous Crude Oils

Abstract:

Application of Aggregation Thermodynamics and Molecular Characterization to Study Asphaltene Precipitation of Bituminous Crude Oils

Md Rashedul Islam, Yifan Hao, Meng Wang, and Chau-Chyun Chen

Department of Chemical Engineering, Texas Tech University

Abstract

Addition of paraffinic feedstock to bituminous crude oil significantly impairs the solvent power of crude oil. The content and nature of paraffins in the blend determines the asphaltene precipitation behavior, i.e. onset point, and yield. Comprehensive thermodynamic modeling offers the necessary tool to understand and track the compatibility of heavy crude oils with paraffinic feedstock, and to predict asphaltene precipitation onset and yield. Recently, an aggregation thermodynamic framework [1] developed from Yen-Mullins hierarchy model for asphaltene [2] has been shown to successfully predict asphaltene precipitation from a wide spectrum of solvent mixtures. Furthermore, a molecule based characterization approach has been utilized to represent Cold Lake bitumen with real hydrocarbon molecules and a refined NRTL-SAC activity coefficient model [4] successfully applied to predict asphaltene precipitation from mixing Cold Lake bitumen with n-alkanes [5]. In this work, this methodology is further demonstrated with two additional heavy oils, i.e., Athabasca and Peace River.

References

[1] M. Wang, Y. Hao, M. R. Islam, and C.-C. Chen, "Aggregation thermodynamics for asphaltene precipitation," AIChE Journal, vol. 62, pp. 1254-1264, 2016.

[2] O. C. Mullins, "The Modified Yen Model†," Energy & Fuels, vol. 24, pp. 2179-2207, 2010.

[3] C.-C. Chen and H. Que, "Method of Characterizing Chemical Composition Of Crude Oil For Petroleum Processing," US Patent Application Number: 2013/0185044, July 18, 2013.

[4] Y. Hao, M.R. Islam, M. Wang, and C.-C. Chen, "A refined Non-Random Two-Liquid segment activity coefficient model for solubility modeling," manuscript in preparation (2016)

[5] M.R. Islam, Y. Hao, M. Wang, T. Kirkes, and C.-C. Chen, "Application of molecular characterization to bituminous crude oil to study asphaltene precipitation," presented at the 2016 AIChE Annual Meeting, San Francisco, CA., November 12-17, 2016.

480408 Hydrate Deposit Formation and Dissociation in Gas-Dominant Systems: Experiments and Modelling Using a High-Pressure Miniloop

Abstract:

The design and operation of subsea oil and gas flowlines requires the capacity to predict the formation and transport of hydrate particles, which remain one of the largest flow assurance concerns due to the short operating period in which they can result in flowline blockage. While significant research has focused on the development of a mechanistic model to describe hydrate blockage in crude oil systems, this behavior has not been well-established in gas-dominant flowlines.

Previous experiments using a single-pass, gas-dominant flowloop operating at high velocities have reported rapid stenosis at the wall supported by the solidification of water droplets entrained in the gas phase and subsequent deposition on the wall. When the same flowloop was operated without liquid entrainment in the gas phase, the rate of hydrate growth was reduced by one order of magnitude and the impact of hydrate formation on frictional pressure drop decreased. While the deposition of hydrate particles from the gas phase on the wall may represent a primary pathway to hydrate blockage formation, fundamental studies have suggested that the growth of a solid hydrate crystal film at the wall may also contribute to stenosis-based blockage mechanism.

To better characterize the growth of a solid hydrate film, a high-pressure, laboratory-scale miniloop was used to characterize the growth rate and morphology of this hydrate film as a function of subcooling and fluid velocity. The miniloop incorporated a specially-designed "deposition test section," which was cooled below the remainder of the flowloop to encourage hydrate film growth. The test section contained four viewing windows in which to visually confirm hydrate formation, a differential pressure transducer, and temperature sensors at the inlet and outlet of the test section to better quantify hydrate deposition. Experiments were conducted using superficial gas and liquid velocities between 0.1 and 1m/s, and test section subcoolings of up to 11 °C. Additionally, hydrate dissociation tests were performed by increasing the bulk fluid temperature, whilst maintaining the temperature of the deposition section; this experimental procedure provided insight into the difficulty of removing an annealed hydrate deposit from the wall.

Preliminary tests were used to investigate hydrate deposition rates on the wall in both the gas and liquid phases. Deposits formed in the gas phase were supported by the condensation and crystallization of water at the wall, which provided an annealing mechanism over time. The formation of a deposit in the liquid phase was limited by the diffusion of gas molecules to the growing deposit, and appeared to be less porous than the deposit formed in the gas phase. The rate of deposit growth in the liquid phase was modelled considering transport limitations for gas molecules. The prediction of hydrate deposit porosity in the gas phase requires additional experimental investigation, as there are currently limited direct experimental studies available to inform this property.

480398 Hydrate Formation Experiments in a Lab-Scale Vertical Flowloop

Abstract:

The formation of gas hydrates in subsea oil and gas flowlines remains a significant operational risk, due to their rapid growth rates and ability to form large aggregates. Much of our understanding of hydrate blockage formation to date has been derived for oil-continuous pipes, where water droplets may be dispersed in the oil phase. However, knowledge of hydrate growth rates and particle transportability in water-continuous systems is less well-understood. In this presentation, we employ the use of a vertical flowloop to investigate hydrate growth behavior using mineral oil and water under steady-state and transient operations.

A high-pressure water tunnel (HPWT), initially constructed to examine hydrate growth behavior on rising gas bubbles following a subsea blowout, was converted for use as a vertical flowloop. Changes included the installation of an ultrasonic flowmeter, differential pressure transducer and viscometer. More recently, the flowloop has been adapted to allow the introduction of mineral oil, as well as the installation of a high-pressure wire feedthrough to measure the conductivity of the flowing fluids, thereby determining the continuous phase.

Experiments were performed using water and mineral oil at water volume fractions of 100, 80, and 50%, at a constant mixture velocity of approximately 1 m/s. Gas addition, hydrate dissociation and flowloop restart tests (with and without hydrates present) were conducted to investigate the effects of hydrate formation under both steady-state and transient operations. A transparent 'bulls-eye' pipe segment, on one of the vertical sections of the loop, allowed for visual observation of hydrate growth behavior. Coupled with a high-speed camera, this allowed for detailed investigation of interfacial phenomena under flow. Contrary to the formation of solid shells on water droplets in oil-continuous systems, we present new evidence suggesting hydrates do not always form complete shells around methane-saturated oil droplets in water-continuous systems. Instead, hydrate crystals were sheared from the gas-water interface soon after formation, and aggregated in the continuous water phase.

These tests have enabled a quantitative comparison of the potential mass transport limitations to hydrate growth, including gas dissolution in the aqueous phase and gas diffusion to the growing hydrate particle. The measured hydrate slurry viscosity increased directly with hydrate volume fraction, increasing the momentum energy required to restart flow following a shut-in. The results from this investigation provide new insight into hydrate growth behavior in high watercut systems, which may be used in the validation of predictive models.

480410 Reduction of Hydrate Film Growth Rate and Cohesive Force By Surface Active Components in Crude Oil

Abstract:

During the production of offshore oil and gas, the cooling of hydrocarbons toward seafloor temperatures enables the formation of gas hydrates, which may restrict fluid flow and ultimately block the flowline. As hydrates aggregate and begin to form a blockage, a hydrate film may grow between individual particles or between hydrate particles and the pipeline wall, respectively resulting in a higher slurry viscosity or a reduced hydraulic diameter. The hydrate cohesive force and hydrate film growth rate are fundamentally controlled by the properties of the interface. While it has been shown previously that hydrate cohesive force may be affected by naturally-occurring surfactants in the crude oil, there is limited data available to inform whether hydrate film growth rate may also be affected by these components. In this study, we have used a micromechanical force (MMF) apparatus to quantify the film growth rate of cyclopentane hydrate at a moderate subcooling. Naturally-occurring surface active species were obtained from two Australian crude oils by solvent extraction, resulting in the separation of three oil fractions: asphaltenes, binding resins, and free resins. Each oil fraction was then added to a chemically inert hydrocarbon phase, and was injected in the MMF experimental cell to quantify its impact on hydrate film growth rate. At mass fractions below 500 ppm, the hydrate film growth rate was reduced by at least one order of magnitude by asphaltenes; both resin classes were less effective at reducing hydrate film growth rate. Despite variations in the film growth measurements, cohesive force trials with the MMF revealed that each of the three fractions was able to reduce the hydrate cohesive force by two orders of magnitude. The presence of each fraction also caused an increase in the wetting angle of the water droplet on the hydrate particle surface, which suggested that these naturally-occurring components may adsorb to the hydrate particle surface.

480411 A New Transient Framework for Assessing Hydrate Formation

Abstract:

As the production of oil and gas moves toward deeper water, the problem of hydrate formation in flowlines becomes more prominent. Traditionally, a strategy of avoidance requiring the addition of large volumes of thermodynamic inhibitors has been pursued. The fiscal constraints of deepwater production may be overcome if hydrates are allowed to form, and the risk of blockage can be successfully managed. This evolving view posits that the flowline may be allowed to operate within the hydrate equilibrium region, while employing both unique operating strategies and novel technologies to avoid blockages. For such methods to be applied with confidence, a robust understanding of both hydrate formation and multiphase transport behaviour are required. The current engineering approach models both concepts independently, where the effect of hydrate growth on slurry transportability is typically isolated from transient flow simulations. In this work, we introduce a new approach to bridge this gap by incorporating the effect of pressure loss due to slurry viscosification and stenosis, or a narrowing of the flow channel, on the driving force for multiphase flow. We present the results of a deepwater flowline case study, in which established hydrate formation and transportability calculations are coupled to a fully transient multiphase flow engine; this unique approach enables flow assurance engineers to evaluate the severity of hydrate formation on a physical basis. For the first time, the quantitative effects of operating within the hydrate stability region can be estimated with respect to hydrate growth, production rate, and fluid properties.

480413 Rheological Characterisation of Hydrate-in-Oil Slurries

Abstract:

Gas hydrates are ice-like solids that may form in crude oil flowlines under high pressure and at low temperature, resulting in high frictional pressure loss and increasing the risk of a non-flowing condition. The viscosity of hydrate slurries has been suggested as a primary criterion to determine blockage risk, but limited rheology data are available to inform the predictive viscosity models. The rheology of hydrate slurries formed from water-in-oil emulsions was measured using a high-pressure controlled-stress rheometer, where gas saturation in the slurry as maintained with a vane-blade rotor. Tests were conducted under isochoric conditions, where the apparent viscosity and pressure of the system were recorded throughout the hydrate growth process, enabling the first comparison of hydrate slurry viscosity as a function of hydrate volume fraction in the liquid phase. The results show that, for 5 to 30 % watercut systems, hydrate formation was repeatedly identified by a significant decrease in pressure and a corresponding increase in slurry viscosity; the maximum and steady-state viscosity values recorded during the hydrate formation process deviated by a maximum of 10% through independent trials. As vane blade rotor was able to maintain saturation of the crude oil phase, these measurements represent the first insight into the worst-case slurry viscosity condition achievable in crude oil flowlines.

Modelling Salts Effect on Hydrate Inhibition with CPA-Electrolyte and Pseudo-Salt Approach with Excess Gibbs Energy Mixing Rules

Xiaohong Zhang, Nuno Pedrosa, Richard Szczepanski
KBC Advanced Technologies Ltd (A Yokogawa Company), London, UK

Abstract

Gas hydrate formation analysis and management have become an essential part of engineering work for the correct design, commissioning and operation of offshore and onshore production and transportation.

The conventional engineering approach for suppression of hydrates formation is to add inhibitors such as methanol or glycol to alter the thermodynamic equilibrium so that hydrates cannot form under production conditions. Although there has been much work recently on low-dosage kinetic inhibitors, there are many situations where it remains essential to use thermodynamic inhibitors. In order to accurately predict the hydrate phase behaviour and the dosage of hydrate inhibitors, the equation of state used has to be able to model systems where hydrocarbons, water and other polar components such as methanol are involved, as well as be able take into account of the presence of salts, and their effect on the solubility of water and hydrate inhibitors, and the accumulated hydrate inhibition.

Traditionally cubic equations of state, such as Peng–Robinson (PR) or Redlich–Kwong–Soave (RKS), with simple mixing rules, are not able to predict correct phase behaviour of highly polar mixtures like water and salts. In order to model not only gas and oil phase but must also an aqueous phase with mixed solvents of alcohols, glycols and salts, either complex mixing rules based on Excess Gibbs energy or additional association terms are needed to provide reliable prediction on complex phase equilibria between the fluid and solid phases. In this study two well established models; CPA (Cubic-Plus-Association) with Electrolyte and the advanced version of RKS with modified NRTL type mixing rule are applied to model the effect of salts and the mixed-solvent electrolytes involving methanol on gas hydrate inhibition.

The study will compare the performance and predictability of the two well-developed models, addressing the importance of accurate fluid model for the prediction of hydrate formation and highlight the best practice on modelling the accumulated effect of salts on hydrate inhibition.

1 Introduction

Gas hydrates are solid ice-like compounds formed by water and stabilised by the presence of "guest" molecules within the hydrate lattice cavities [1]. As these guest molecules are typically light gases and hydrocarbons, hydrates are of particular importance to the oil and gas industry. Although hydrates may be an important future hydrocarbon energy resource with the development of appropriate means of capture, transportation and storage, their presence in oil and gas exploration and production is a major challenge due to the operational and safety concerns. Hydrates can exist in-situ in petroleum reservoirs where they can cause blowouts if drilled into inadvertently[2]. Alternatively, extreme conditions of temperature and pressure (typically low temperatures and high pressure) mean that hydrates can easily form during the drilling process if fluids containing water, drilling muds, seawater or produced water, come into contact with reservoir fluids[3,4]. Formation of solid hydrates can also cause blockages in subsea risers, blow out preventers (BOPs), choke and kill flowing lines. Conditions during well shut-in are particular in favour of hydrate formation if high pressures are combined with falling temperatures[1].

To prevent and suppress the gas hydrate formation in oil and gas exploration and production, the conventional engineering approach is to use hydrate inhibitors such as methanol or glycols (MEG, DEG, TEG) to alter the thermodynamic properties of the aqueous phase so that hydrates cannot form under the production conditions[5]. During the production, the effect of the presence of salts in produced or sea water must also be considered. This work will show that their effect cannot be neglected. Use of thermodynamic inhibitors, however, is expensive and can lead to high capital expenditure for storage and significant operational costs. Thus it is desirable to understand the mechanism of thermodynamic hydrate inhibition so that correct operation schemes to avoid over-dosage can be implemented to control and prevent hydrate formation economically.

To have a correct operation scheme and avoid over-dosage, thermodynamic modelling is essential for helping understanding the phase behaviour of the fluid and the inhibitors, and most importantly the effect of the additional salts on hydrate inhibition as these ionic salts can act as hydrate inhibitors and are always present in seawater, formation and produced water under the drilling and production conditions. In this work, two well-developed thermodynamic models in Multiflash^{TM} are applied to model the effect of ionic salts on the hydrate inhibition; the CPA (Cubic-Plus-Association) with Electrolyte and the advanced version of RKS with NRTL type mixing rule (RKSA-Infochem)[6,7].

2 Modelling Gas Hydrates and Hydrate Inhibition

2.1 Gas Hydrates Structures

The driving force of gas hydrate formation depends on the three key criteria; the presence of water, hydrate forming components and the right conditions of pressure and temperature. Gas hydrates form when the criteria are satisfied and water forms a cage-like structure around a "guest" molecule. There are many compounds that can form hydrates and several hydrate structures. However, in the oil and gas industry only three forms of hydrate structure need to be considered; Structures sI, sII and sH[1]. Among those, the structure I hydrate is the simplest structure with two types of cavities, the components that are found to form structure I hydrates in oil and gas production and processing are methane, CO_2, H_2S and ethane. Similar to structure I hydrate, the structure II hydrate also has two types of cavities but is more complex and much larger, therefore large molecules such as propane and iso-butane can fit into the cavities of the hydrate structures to form structure II hydrates[8,9]. The structure sH is the last hydrate structure that is relevant to the oil and gas industry. Unlike Structures I and II, structure H is never formed by a single pure component but requires two hydrate formers, a small molecule such as methane and a larger Type H forming molecule. Typical of the latter are branched butanes and cyclics. Structure H is made up from three types of cage: a regular dodecahedron, an irregular dodecahedron and an irregular 20 sided polyhedron[10].

The larger molecules of a gas mixture normally determine the hydrate structure formed so the presence of propane and i-butane in natural gas mixtures means that the normal form found in gas production and processing is Structure II hydrate (sII).

2.2 Gas Hydrate Forming Components

Gas hydrate formation requires small components to occupy the hydrate lattice to stabilise the hydrate structures. As one knows, many of the components in natural gas, gas condensates or oils can form gas hydrates but not all of them are not explicitly identified in compositional analyses and therefore will not be included in any modelling of hydrate formation. To develop a hydrate model for a typical component, the relevant hydrate dissociation data are required for establishing the model parameters[1]. However, the most common hydrate forming components relevant to oil and gas production and processing are well-understood and can be also modelled by the PVT software, Multiflash^{TM}. The detailed list of the hydrate forming components is summarised in Table 1, where the components marked with star are large molecules, requiring small molecules like methane to help forming hydrate structures.

Natural gases	Condensates/oils	Process
Nitrogen	Benzene*	Ethylene
Carbon dioxide	Cyclopentane*	Propylene
Hydrogen sulphide	Cyclohexane*	Other olefins
Methane	Methylcyclopentane*	
Ethane	Cycloheptane*	
Propane	Methylcyclohexane*	
Butane*	Ethylcyclopentane*	
Isobutane	Cyclooctane*	
Neopentane*	1,1 Dimethylcyclohexane*	
	cis 1,2 Dimethylcyclohexane*	
	2 Methyl butane*	
	2,2 Dimethyl butane*	
	2,3 Dimethyl butane*	
	2,2,3 Trimethyl butane*	
	3,3 Dimethyl pentane*	
	2,2 Dimethyl pentane*	
	2,3 Dimethyl-1-butene*	
	3,3 Dimethyl -1-butene*	
	3,3 Dimethyl-1-butyne*	
	cis cyclooctene*	
	bicyclo[2,2,2] oct-2-ene*	
	Adamantane*	

Table 1 Hydrate forming components. *helper gas needed to form hydrates

2.3 *Gas Hydrates and Hydrate Inhibition Calculation*

The calculation of phase equilibria involving hydrates is, in principle, no different from any other phase equilibrium calculation. The phase equilibrium equations must be solved to satisfy the equality of fugacities of each component in every phase:

$$f_{i,1} = f_{i,2} = f_{i,3} = \ldots = f_{i,p} \tag{1}$$

For a multi-component system, solving the equation (1) is accurately very challenging and requires a robust and complex phase equilibrium algorithm involving not only gas hydrate phases but also other fluid and solid phases[6]. Usually it is necessary to consider up to six phases: gas, hydrocarbon liquid, aqueous liquid, ice, Hydrate I and Hydrate II. To model the effect of salts on hydrate inhibition, additional solid salt phases are also required, leading up to total of 17 phases that may be considered. The types of models typically used to describe the properties of each of these phases in Multiflash^{TM} are briefly described in the following sections.

2.3.1 Modelling Gas Hydrates

The gas hydrate model developed in Multiflash^{TM} is based on the work of Van der Waals and Platteeuw[11,12]. It is assumed that the formation of gas hydrate is thermodynamically as having two steps: (1), pure liquid water solidifies to form an empty hydrate lattice; (2), gas molecules are adsorbed into the cavities in the hydrate lattice. The former is analogous to the formation of ice and is described by the same thermodynamic equations, although the parameter values of hydrate lattice are different. The latter is described by the statistical mechanics of the hydrate adsorption process that was derived by Van der Waals and Platteeuw for modelling the interaction between the gas molecules and the hydrate lattice using an effective intermolecular potential. The general mathematic formulation for describing the change of the difference in chemical potential between an empty and filled hydrate lattice is given by[1,6]

$$\frac{\Delta\mu^H}{RT} = \sum_k v_k ln\left(1 - \sum_i n_{i,k}\right) \tag{2}$$

where the first summation is over the number of cavities k in hydrate lattices, and the second one is the summation over the component, *i*. The $n_{i,k}$ is the probability that a cavity, *k*, is occupied by a gas or light hydrocarbon molecule, *i*, which is defined by

$$n_{i,k} = \frac{C_{i,k} f_i}{1 + \sum_j C_{j,k} f_j} \tag{3}$$

The f_i is the fugacity of the component, *i*, that is calculated by thermodynamic fluid models such as CPA and RKSA-Infochem. $C_{i,k}$ is the Langmuir constant of the component, *i*, which is cavity specific, temperature-dependent property.

$$C_{i,k} = C_{i,k}(T, \boldsymbol{a}_{i,k}) \tag{4}$$

where the adjustable parameters $a_{i,k}$ are estimated by matching experimental hydrate dissociation data. This approach has been widely used and tested for many years and is capable of accurately representing the experimental data on hydrates. The gas hydrate model in MultiflashTM has been developed with large sets of experimental data including proprietary data from industrial gas hydrate projects and the data from public domain[1]. Later the model was extended and further improved as part of a deepwater JIP project[13]. In general the model can give accurate prediction of hydrate dissociation temperature within one degree. The examples of the performance of the models are shown in Figure 1 and Figure 2.

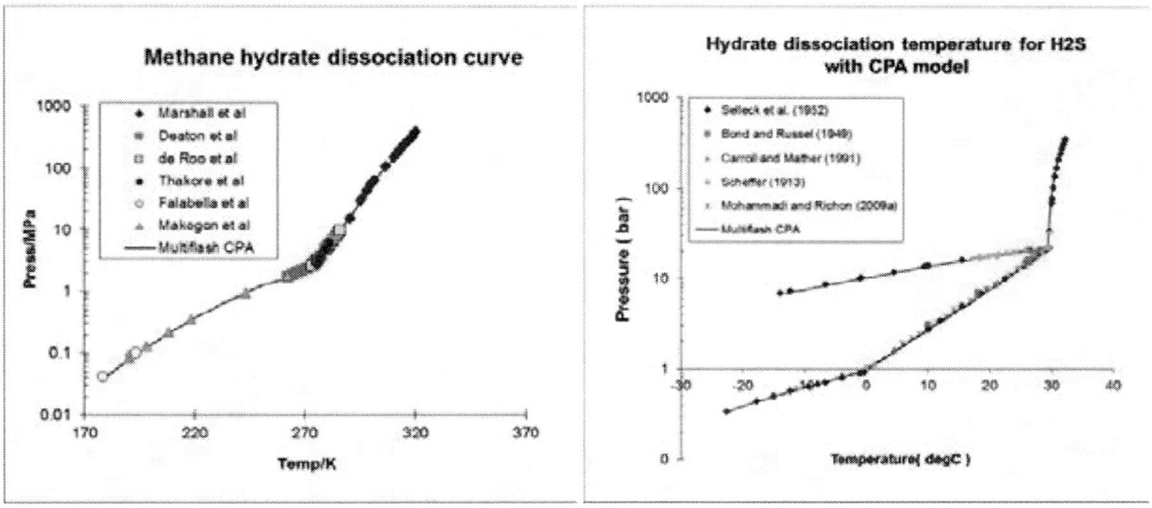

Figure 1 Examples of the Gas Hydrate dissociation curves of Methane and H_2S predicted by Multiflash™[1]

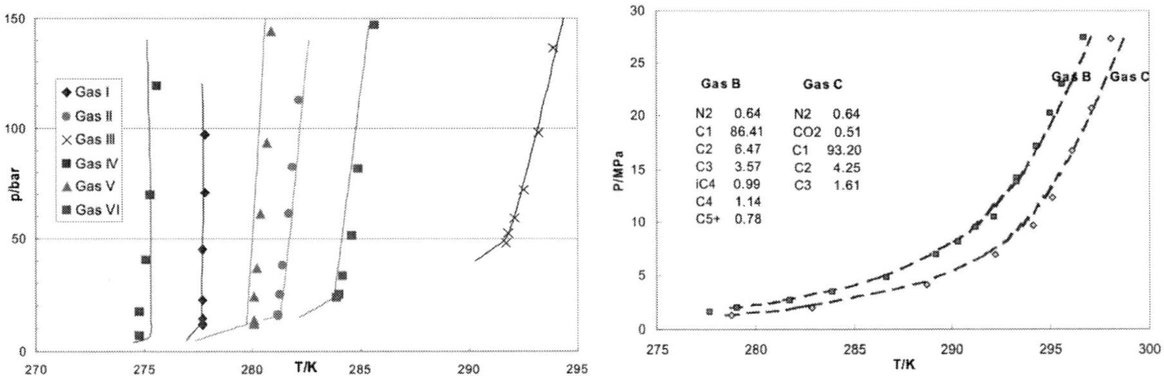

Figure 2 Examples of the Gas Hydrate dissociation curves of natural gases and condensates predicted by Multiflash™[1]

2.3.2 Fluid Phases

As described in equations (2) and (3), the fluid models play an important role in gas hydrate calculations where thermodynamic fluid models provide the fugacity calculations of the hydrate forming components as well as for the fluid phases[1,6,14]. Thus modelling the gas hydrate phase behaviour accurately demands accurate modelling for not only gas and oil phase, but must also the aqueous phase that will usually contain alcohols, glycols, electrolytes and several possible solid phases.

The general fluid modelling approach in Multiflash™ is to enhance the basic cubic equation of state (EOS) by adding appropriate thermodynamic terms to the total residual Helmholtz free energy. Hence the general fluid phase model in Multiflash™ has the following structure

$$A^{res} = A^{res}_{RKS} + A_{assoc} + A_{electrolyte} \tag{5}$$

where A^{res} is the total residual Helmholtz energy of the fluid phases, A_{assoc} is the contribution from chemical association and $A_{electrolyte}$ is the term that accounts for contributions of electrolyte solutions. A^{res}_{RKS} is the residual term described by the cubic EoS, e.g. Redlich-Kwong-Soave EoS (RKS).

The aqueous phase is perhaps the most difficult to represent, particularly if inhibitors such as methanol and/or salts are present. Having a reliable model to accurately describe the aqueous phase behaviour is crucial for modelling gas hydrates. However, the aqueous phase cannot be adequately modelled with a conventional cubic EOS. Multiflash™ offers the advanced version of RKS (RKSA) model either with an NRTL type mixing rule (RKSA-Infochem) or with an explicit association term (CPA)[14]. The common cubic EoS A^{res}_{RKS} term based on RKS can be expressed by

$$A^{res}_{RKS} = -NRTln\left(1 - \frac{b}{V}\right) - N\frac{a}{b}ln\left(1 + \frac{b}{V}\right) \tag{6}$$

For the RKSA-Infochem approach, only the first term in equation (5) is considered that includes the "a" parameters determined by fitting the vapour pressure to improve the saturation pressure prediction and the Peneloux volume correction is also added to improve the liquid density[7]. The modified NRTL-Infochem mixing rule then enables the RKSA to model the phase behaviour of the polar components such as water, methanol and glycols. The form of the modified NRTL Infochem mixing rule is:

$$a = b\sum_i n_i \frac{\sum_j n_j b_j \xi_{ji} G_{ji}}{\sum_j n_j b_j G_{ji}} \tag{7}$$

Where

$$\xi_{ji} = \frac{2\sqrt{a_i a_j}(1 - k_{ji})}{b_i + b_j} \tag{8}$$

And

$$G_{ji} = exp\left(\frac{\alpha_{ji}(\xi_{ji} - \xi_{ii})}{RT}\right) \tag{9}$$

In general $k_{ji} \neq k_{ij}$for highly polar components and $\alpha_{ji} = \alpha_{ij}$. The advantage of this approach is that if α_{ij} is set to zero, the mixing rule reduces to the conventional Van der Waals mixing rule that is normally used with the standard cubic EoS.

$$a = \sum_{ij} n_i n_j \sqrt{a_i a_j}(1 - k_{ji}) \tag{10}$$

Where the binary interaction parameters are now symmetric, $k_{ji} = k_{ij}$.

With the binary interaction parameters (BIPs), this NRTL-Infochem mixing rule enables the cubic EoS to handle the properties of hydrate inhibitions in a thermodynamically consistent way allowing all the effects of adding inhibitors to be modelled including the depression of hydrate points, the depression of the ice formation and the reduction in the vapour pressure of water.

In Multiflash™, the chemical association approach (CPA) based on the Advanced version of RKS (RKSA) is the recommended model for gas hydrate formation and inhibition modelling. The general numerical formulation of CPA in Multiflash™ is based on the work of Kontogeorgis et al [14], which has the form:

$$A_{assoc} = \sum_i n_i \sum_{A_i} \left[lnX_{A_i} - \frac{1}{2} X_{A_i} + \frac{1}{2} \right] \tag{11}$$

Where α_{ij} is the mole number of component, i, and X_{A_i} is the mole fraction of molecules not bonded at the association site A_i, given by

$$\frac{1}{X_{A_i}} = 1 + \rho \sum_j n_j \sum_{B_j} X_{B_j} \Delta_{A_i B_j} \tag{12}$$

Where $\Delta_{A_i B_j}$, the association strength is expressed as

$$\Delta_{A_i B_j} = b_{ij} \beta_{A_i B_j} \, g(\rho) \left[exp\left(\frac{\varepsilon_{A_i B_j}}{RT} \right) - 1 \right] \tag{13}$$

Where $g(\rho)$ is the radial distribution function. The association energy $\varepsilon_{A_i B_j}$ and the association volume $\beta_{A_i B_j}$ are the two parameters that are determined by fitting the experimental VLE and thermal data. For the details of the CPA model, refer to the work by Kontogeorgis et al [14].

In general both the RKSA-Infochem and CPA models are capable of giving accurate predictions on the fluid and solid phases. But CPA offers better prediction of the methanol partitioning and more reliable extrapolation performance in the region outside of the conditions covered by the experimental data.

2.3.3 Salinity Effect on Gas Hydrates Formation

Salts, as a natural hydrate inhibitor, can be as effective as methanol and glycols on hydrate inhibition by reducing the water activity in aqueous solution to supress the hydrate formation[17-19]. Thus modelling salt inhibition on gas hydrate formation accurately is important for the oil and gas exploration and production as the formation water and drilling muds may contain concentrated solutions of sodium chloride, calcium chloride or other salts.

The electrochemistry of aqueous salt solutions is complex but for the main objective of modelling salts effect on hydrate inhibition, it is reasonable to simplify the complexity, at the same time, and capture the key features of the salts effect on the properties of the fluid and the hydrates phase behaviour. Thus in order to model the salinity effect on hydrate formation accurately, the model must accurately predict: (1): salts stay in aqueous phase; (2), salts precipitate when solubility limit is reached; (3), depression of hydrate and ice formation points; (4), salting-out effect; (5), presence of salts reduces water vapour pressure as well as ensure the approach to be thermodynamically consistent with the gas hydrate model and the fluid models. In Multiflash^{TM}, there are two practical approaches available that can be applied to model the effect of salts on hydrate formation; one is the conventional electrolyte approach with CPA and another one is a pseudo equivalent-NaCl salt component approach with RKSA-Infochem[7].

The electrolyte approach with CPA in Multiflash^{TM} is based on the Debye-Hückel theory plus virial corrections and Born term to predict the effect of salts on hydrate inhibition[14-16]. The general electrolytes contribution to the total residual Helmholtz free energy is:

$$A_{electrolyte} = A_{DH} + A_{VR} + A_{Born} \tag{14}$$

Where A_{DH}, the Debye-Hückel describes the electrostatic ion-ion interactions. A_{Born} for the interactions between ions and the surrounding medium, and A_{VR}, the virial terms for the short range interactions between the ions. The Debye-Hückel term, the contribution of the electrostatic interactions between charged ions to the total residual Helmholtz free energy can be written in the following form:

$$A_{DH} = -\frac{4A_0V}{Z^+Z^-}\sum_{i,j} n_i n_j z_i z_j \frac{\tau_{ij}}{\left(b_{ij}^0\right)^3} \tag{15}$$

Where A_0 is a constant, Z^+ and Z^- are the total positive and negative charge respectively. The index i represents the cations, and j the anions, and z_i is the charge of the ions. b_{ij}^0 is specific Debye-Hückel parameter for each pair of ions and τ_{ij} is defined as:

$$\tau_{ij} = ln(1 + y_{ij}) - y_{ij} + 0.5\, y_{ij}^2 \tag{16}$$

and

$$y_{ij} = \frac{b_{ij}^0}{\sqrt{V\epsilon T}}\Phi^{1/2} \tag{17}$$

where V is the total volume, T is the temperature, ϵ the relative permittivity of the solvent and Φ the ionic strength defined as

$$\Phi = \frac{1}{2}\left(\sum_{i\in cation} n_i z_i^2 + \sum_{j\in anion} n_j z_j^2\right) \tag{18}$$

In this electrolyte approach, it is assumed that the salts are fully dissociated into cation and anion ions in aqueous solution. The presence of the charged ions causes the electrolyte solution to deviate much more from ideal solution behaviour than a non-electrolyte solution does. For simple electrolytes, water is the only solvent and the electrolytes are solute. In order to model the effect of salts on hydrate inhibition in oil and gas production environment, however, the aqueous solution usually contains other species like water, methanol, glycols and salts as well as other dissolved acidic gases. Therefore the interactions between ions and the surrounding solvents have to be described by "mixed solvents" or pseudo solvents, and such interactions can be expressed by Born equation for describing the contribution to the total residual Helmholtz free energy in terms of the presence of polarizable solvents affecting the relative permittivity of the dielectric medium. In the implementation of Multiflash™ CPA-Electrolyte, the second and third virial contributions are also considered. For the details of Born equation and virial corrections, refer to the scientific textbooks[14-16].

With the additional virial corrections and Born term, the model is able to accurately predict the suppressing effect of the mixed solvents (water, methanol and glycols and salts) on hydrate formation, the solubility of gases in salted water, the freezing point depression and vapour pressure lowering, see in Figure 3.

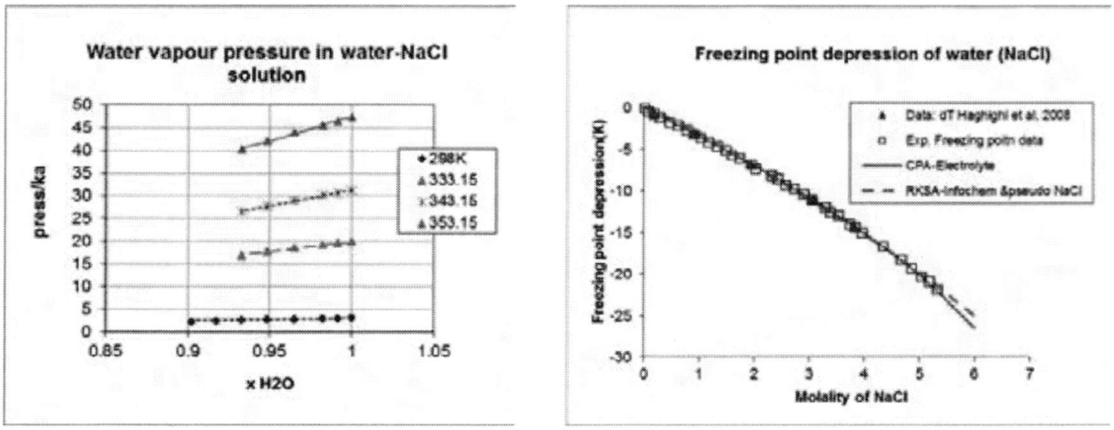

Figure 3 Effect of NaCl on water vapour pressure and freezing temperature, predicted by Multiflash™[17,18].

On the other hand, the pseudo-salt component approach based on RKSA-Infochem follows the cubic EoS framework by assuming that the salts are represented by a single pseudo NaCl-equivalent salt component (pseudo-NaCl) with physical properties and binary interaction parameters that ensure the model captures the key features of the salts effect on ice, hydrate formation and the properties of fluid phases. The hypothetic pseudo salt component (pseudo-NaCl) behaves like a typical non-volatile and heavy component with high critical temperature and low critical pressure (991.7K,1.86MPa), staying in the aqueous liquid phase or precipitating as a solid when its solubility limit is reached. Setting the interaction parameters of the pseudo-NaCl with hydrocarbons and light gases to a suitable large value also ensures that the salt component, pseudo-NaCl will not enter the hydrocarbon liquid phase.

For salt mixtures containing several types of salts, such as formation water and produced water or brines, rules are required to convert the mixed salt solution into a single equivalent-NaCl solution so that the salinity effect can be modelled by a single pseudo-NaCl[7]. This conversion assumes that the reduction of the water activity by the mixed salts is the same as that by the single pseudo-NaCl in the aqueous solution. Two typical examples of the salt composition of Forties formation water and the North Sea brine are given in Table 2, and the equivalent amount of the pseudo-NaCl for representing the mixed salts is also included in the table.

Salt names	North Sea Brine, wt%	Forties Formation Water, wt%	Synthetic Formation Water, wt%
NaCl	2.354	6.993	12.06
$CaCl_2$	0.116	0.735	1.04
$MgCl_2$	0.524	0.186	0.12
KCl	0.086	0.066	0.56
Na_2SO4	0.428	-	-
$SrCl_2$	-	0.099	0.04
$BaCl_2$	-	0.036	0.01
Equivalent pseudo-NaCl for 100g of water	3.362g	8.637g	15.80g

Table 2 Examples of salts analysis[1]

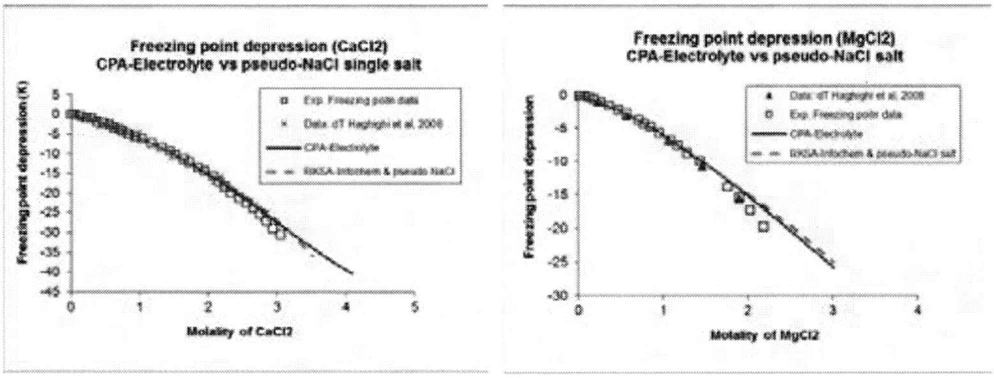

Figure 4 Freezing point depression of water with $CaCl_2$ and $MgCl_2$, predicted by Multiflash™[18]

The Figure 4 shows the freezing point depression of water as a function of the molality of $CaCl_2$ and $MgCl_2$ respectively, are well predicted by both the CPA-Electrolyte and pseudo-NaCl salt component approach with RKSA-Infochem model.

3 Results and Discussion

In this section, both the CPA-Electrolyte and pseudo-NaCl with RKSA-Infochem model are applied in the investigations of the salinity effect on hydrate formation for several types of mixtures, ranging from simple binary systems to a natural gas, gas condensate and black oil.

Before the calculations, the salinity analysis is necessary to obtain the relevant amounts of ions and the single pseudo-NaCl salt component to the relative amount of water in the mixtures.

3.1 Methane Hydrate Formation in Aqueous Mixed Electrolyte Solutions

The methane structure I hydrate has been investigated for over a century and many experimental measurements have been carried out, covering a wide range of temperature and pressure as shown in Figure 1[1]. To investigate the effect of salts and mixed electrolyte solutions on methane hydrate formation, the data measured by Dholabhai[19] are considered. The results of the hydrate dissociation curves predicted by Multiflash are presented in three groups; NaCl and KCl mixed electrolytes, NaCl and $CaCl_2$ mixed electrolytes, and NaCl, KCl and $CaCl_2$ mixed electrolytes. The results are summarised in Figure 5, Figure 6 and Figure 7 respectively.

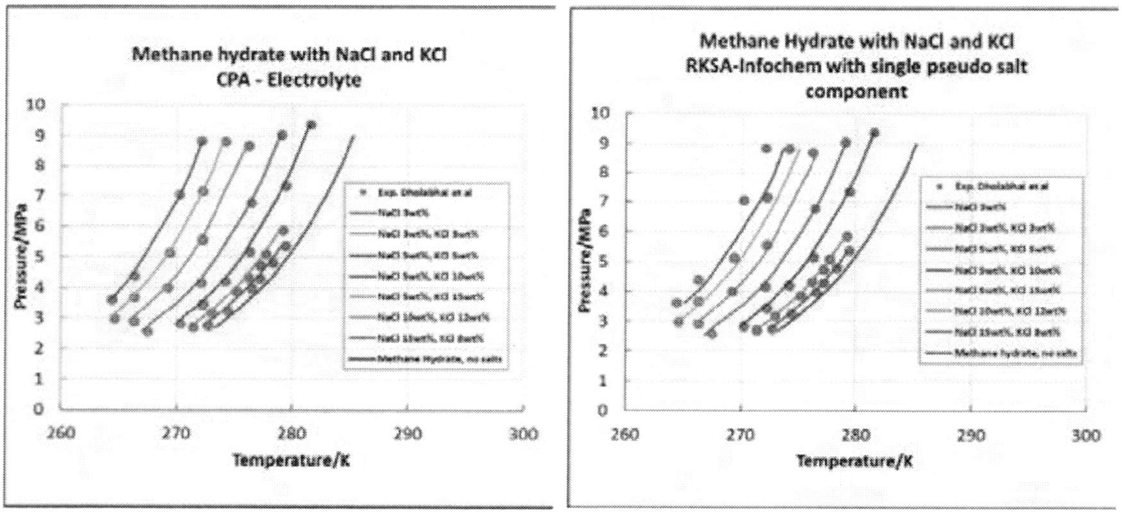

(a) (b)

Figure 5 Methane hydrate dissociation curves with NaCl and KCl. (a), solid lines are predicted by CPA-Electrolyte in Multiflash. (b), solid lines are calculated by RKSA-Infochem with a single pseudo-NaCl salt component.

As clearly shown in Figure 5 (a), the suppression by the mixed electrolytes of NaCl and KCl on hydrate dissociation curves are well predicted by the Multiflash CPA-Electrolyte model, the accuracy of the dissociation temperatures predicted is within 0.5 degree. It is very interesting to see that the sodium chloride by weight is more effective than that of potassium chloride to suppress the hydrate formation. Similar results are also obtained with the single pseudo-NaCl approach, and the predicted hydrate dissociation curves are consistent with the experimental data up to 20wt% of the total salt concentration. However with increasing the salt concentration over 20wt%, the uncertainty of the predicted hydrate dissociation temperatures tends to increase from 0.5 degree up to ~1.5 degree.

For the mixed electrolytes containing NaCl and $CaCl_2$, the predicted results are presented in Figure 6 together with the experimental data. As shown in Figure 6, both models give the consistent and accurate predictions of the effect of salts on hydrate formation as compared with the experimental data. This demonstrates that not only the CPA-Electrolyte model but also the approximated single pseudo-NaCl approach is capable of representing the mixed salts effect on hydrate formation with the total salt concentration up to 21 wt%. The results also show that the effect of the mixed salts with NaCl 6wt% and $CaCl_2$ 10wt% gives the same effect on suppressing hydrate formation as that of NaCl 10wt% and $CaCl_2$ 6wt%.

For the three-salt mixed electrolytes with NaCl, KCl and $CaCl_2$, the CPA-Electrolyte model is slightly better than that of the single pseudo-NaCl approach as shown in Figure 7.

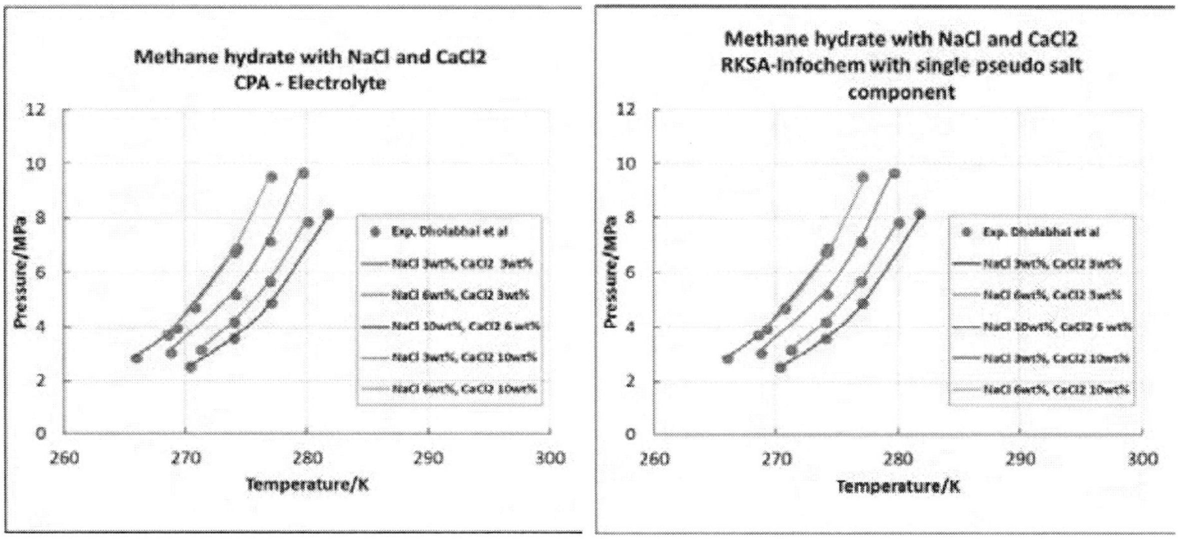

(a) (b)

Figure 6 Methane hydrate dissociation curves with NaCl and $CaCl_2$. (a), solid lines are predicted by CPA-Electrolyte in Multiflash™. (b), solid lines are calculated by RKSA-Infochem with a single pseudo-NaCl salt component.

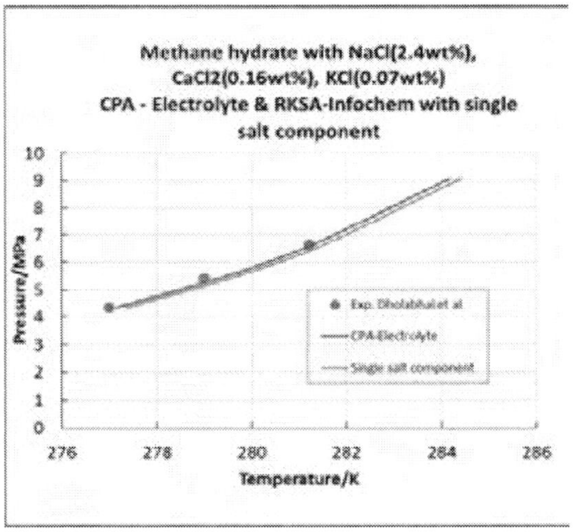

Figure 7 Methane hydrate dissociation curves with NaCl, KCl and $CaCl_2$.

3.2 Methane-CO_2 with Single/Mixed Electrolyte Solutions

A binary methane-CO_2 gas mixture with single and mixed electrolyte solutions is considered to validate the two models. The data of hydrate dissociation curves with single and mixed electrolyte solutions by Dholabhai *et al*[20] were used in the study. The concentrations of methane and CO_2 are fixed to be 80mol% and 20mol% in salt free gas mixture respectively. The single electrolytes are NaCl, KCl and $CaCl_2$ with concentrations ranging from 5wt% to 20wt%, whereas the mixed electrolyte consists of NaCl 5wt%, KCl 5wt% and $CaCl_2$ 10wt%.

The phase behaviour of methane-CO_2-water mixtures is complex involving vapour, hydrocarbon liquid, liquid water, ice, hydrates. Additional solid phases for salts are also needed when the salts are added to the systems. The VLE and the hydrate dissociation curves predicted by Multiflash CPA are presented in Figure 8. As shown in the figures, the phase boundary of both the VLE and hydrate dissociation curves depends on the composition of the methane and CO_2 in the mixture. Both methane and CO_2 forms structure I hydrate but the size, the stoichiometric ratio (number of water molecules to the number of gas molecules) between the methane and CO_2 hydrate is different, leading to different hydrate phase behaviour with temperature and pressure.

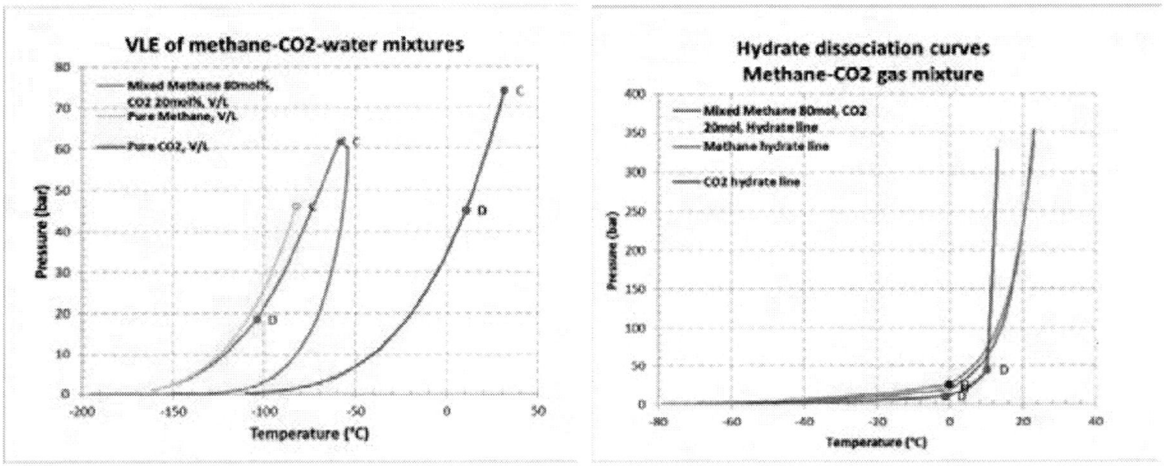

Figure 8 The VLE and hydrate dissociation curves of Methane-CO_2-Water mixtures, predicted by Multiflash CPA.

The hydrate dissociation curves of the methane-CO_2 gas mixtures with salts predicted by Multiflash are summarised in Figure 9 to Figure 12 respectively. For all the cases, the CPA-Electrolyte model gives very accurate predictions of the suppression effect of salts on the hydrate dissociation curves with uncertainty of the dissociation temperature within 0.5 degree.

For the RKSA-Infochem plus a single pseudo-NaCl approach, the hydrate dissociation curves predicted by the model is quite consistent with the experimental data with average uncertainty of dissociation temperature within 1.0 degree. However for the case with NaCl 20 wt% and the case with the mixed salts in Figure 12, the uncertainty of the dissociation temperature predicted by the model is within 1.7 degree and 1.0 degree respectively.

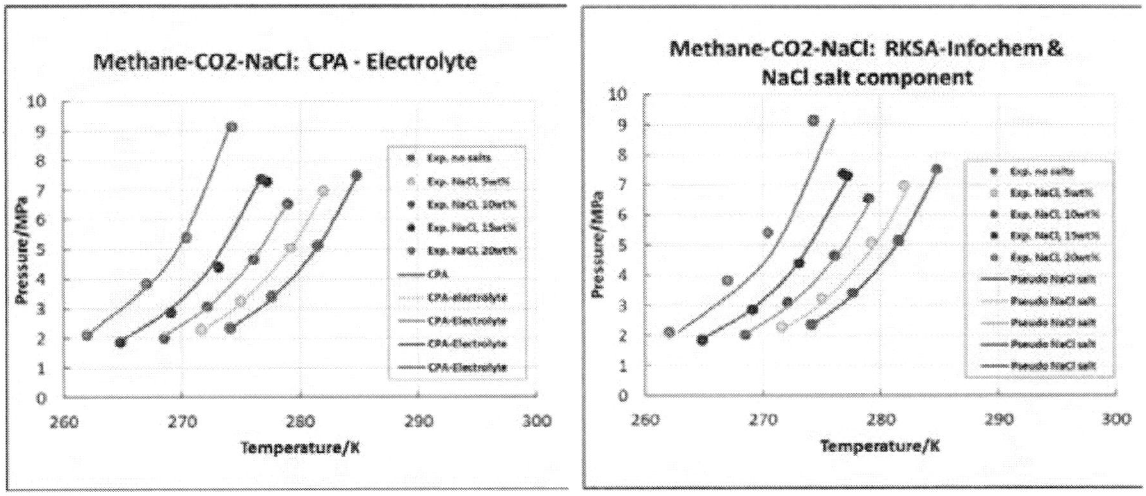

Figure 9 Hydrate dissociation curves of methane-CO_2 with NaCl.

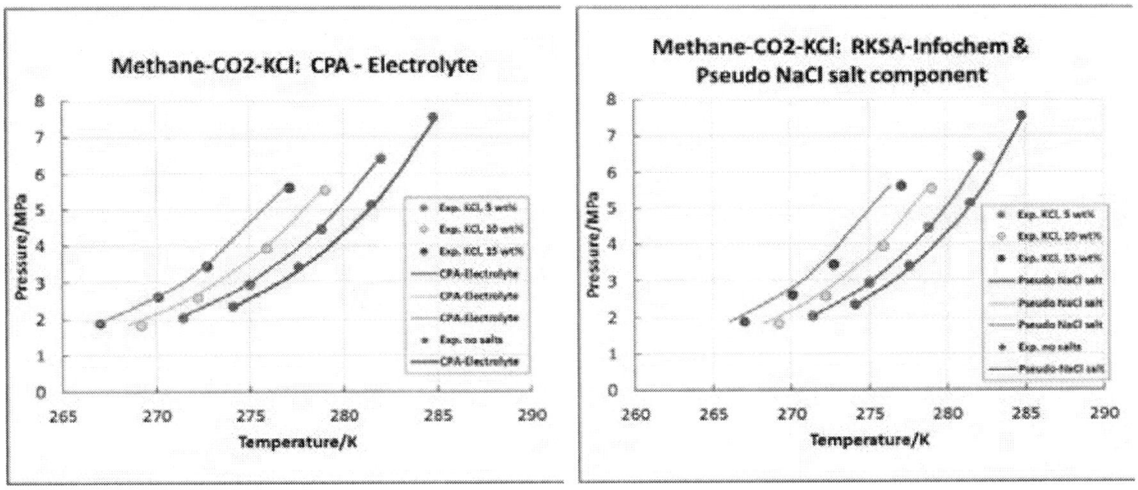

Figure 10 Hydrate dissociation curves of methane-CO_2 with KCl.

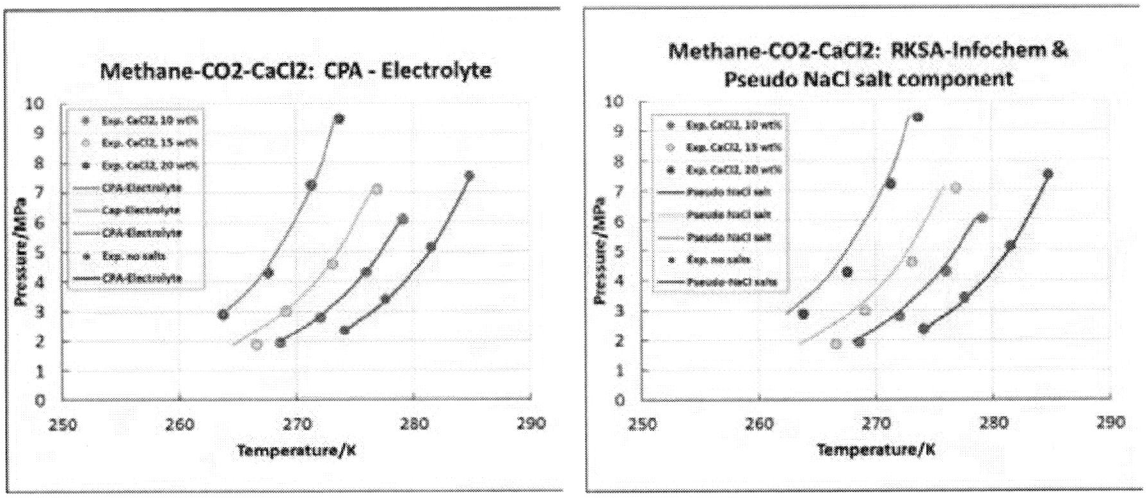

Figure 11 Hydrate dissociation curves of methane-CO_2 with $CaCl_2$.

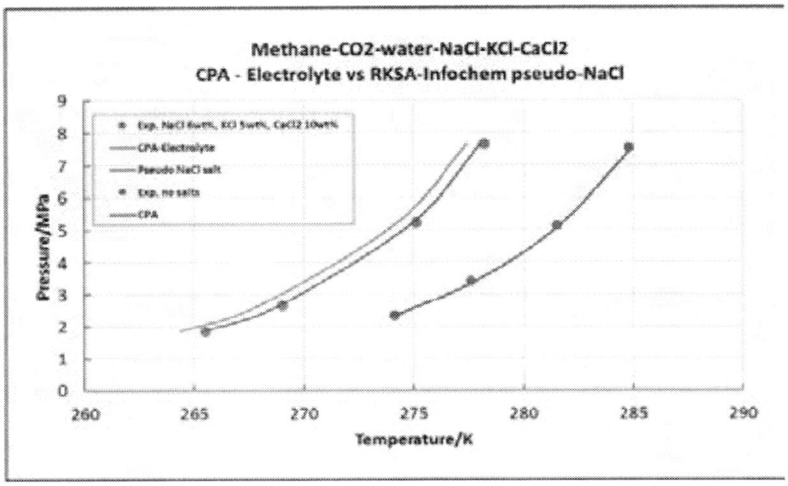

Figure 12 Hydrate dissociation curves of methane-CO_2 with NaCl, KCl and $CaCl_2$.

3.3 Methane-CO_2 with Electrolytes and Methanol

The methane-CO_2 gas mixture with electrolytes and methanol is also applied to validate the two models. The data of hydrate dissociation curves are cited from the work by Dholabhai *et al*[21].

The main objective of this evaluation is to check how the mixed solvents of electrolytes and methanol affect the hydrate formation. Both methanol and the electrolyte concentration of NaCl, KCl and $CaCl_2$ used in the study are 5wt%, 10wt% and 15wt% respectively. One mixed electrolytes of NaCl, KCl and $CaCl_2$ with 5wt% each is also studied. The results of all the cases are summarised in Figure 13 to Figure 15.

As shown in Figure 13 to Figure 15, the hydrate dissociation curves predicted by the CPA-Electrolyte model are in very good agreement with the experimental data for all the cases, whereas the RKSA-Infochem with pseudo-NaCl model works well with lower concentrations of electrolytes and methanol. However with increasing the concentration of the electrolyte and methanol, the hydrate dissociation curves are slightly over-predicted by the model.

Another interesting finding from this study is that the effect of single NaCl and $CaCl_2$ electrolyte on hydrate formation by weight is actually much stronger than that of methanol.

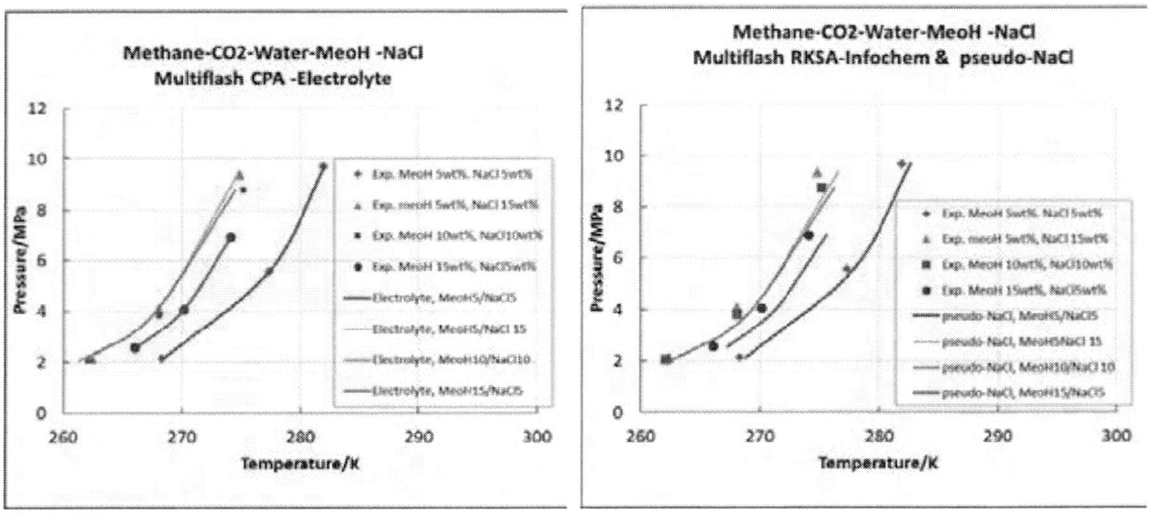

Figure 13 Hydrate dissociation curves of methane-CO_2 with NaCl and methanol.

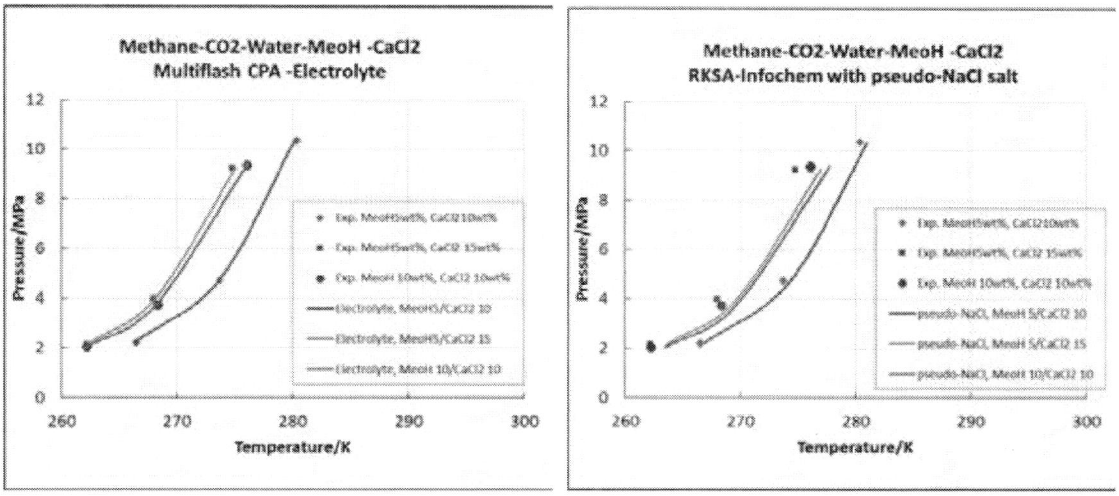

Figure 14 Hydrate dissociation curves of methane-CO_2 with $CaCl_2$ and methanol.

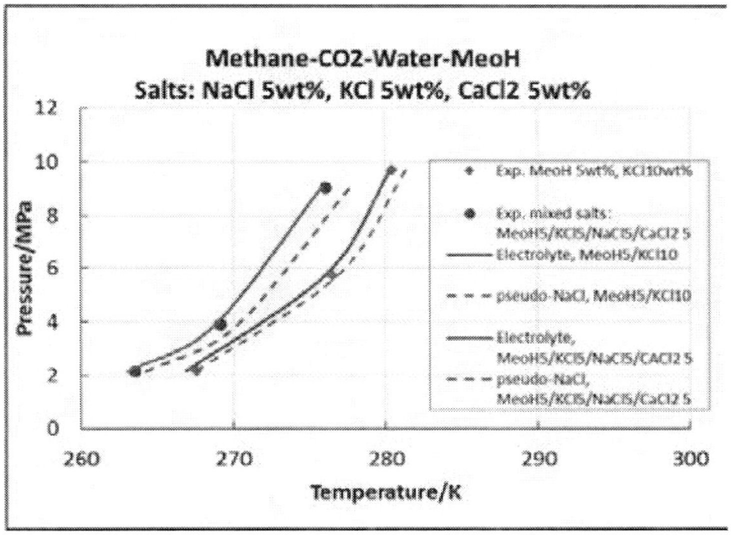

Figure 15 Hydrate dissociation curves of methane-CO_2 with mixed electrolytes and methanol.

3.4 Natural Gas, Gas Condensate and Black Oil with Formation Water

The study on the effect of salts on hydrate formation is now extended to a natural gas, gas condensate and black oil, which are summarised as follows:

- Natural gas with Forties Formation Water (FFW)[22]
- Gas condensate with Synthetic Formation Water (SFW)[23]
- Black oil with mixed electrolytes ($CaCl_2$, $MgCl_2$) and FFW[24]

The three cases are cited from the work by Tohidi *et al*[22-24]. The details of the fluid compositions of the gas and gas condensate are summarised in Table 3, and the composition of the black oil in Table 4. The salt analysis of the Forties formation water and synthetic formation water can be found in Table 2. For mixtures of gas condensate and black oil, a standard fluid characterisation procedure in Multiflash is applied to define all the physical properties of the petroleum fractions required by the thermodynamic fluid models[25].

The hydrate dissociation curves of the natural gas and gas condensate with and without FFW and SFW are presented in Figure 16. The figures clearly show that the predicted hydrate dissociation curves by the two models are in very good agreement with the experimental data. The results also show that the suppression effect by the FFW and SFW on hydrate formation is quite significant, 2 to 3 degree for the natural gas and 2 to 5 degree for the gas condensate, depending on the pressures.

For the black oil, the results are presented in Figure 17, in which the mixed salts A is a mixture of NaCl 6.01wt%, $CaCl_2$ 1.49 wt% and $MgCl_2$ 0.19wt%, whereas the mixed salt B consists of NaCl 8.46 wt%, $CaCl_2$ 3.04wt% and $MgCl_2$ 0.87 wt%. Both models can give accurately prediction on the effect of mixed electrolytes on the hydrate dissociation curves for the black oil. Since the results by the pseudo-NaCl with RKSA-Infochem model is almost the same as that of CPA-Electrolyte, the results are not presented in Figure 17. Based on the results, it is interesting to know that the mixed salt A gives the same effect as the Forties formation water on the hydrate formation.

Components	Natural Gas (mol%)	Gas Condensate (mol%)
Methane	85.93	68.73
Ethane	6.75	9.88
Propane	3.13	4.22
i-Butane	0.71	0.48
n-Butane	0.88	1.49
i-Pentane	-	0.32
n-Pentane	0.57	0.86
C6	-	1.00
C7	-	2.02
C8	-	2.33
C9	-	1.78
C10	-	1.25
C11	-	0.89
C12	-	0.64

C13	-	0.52
C14	-	0.65
C15	-	0.48
C16	-	0.36
C17	-	0.24
C18	-	0.25
C19	-	0.20
C20	-	0.78
CO2	1.31	0.27
N2	0.72	0.36

Table 3 The compositional analysis of the natural gas and gas condensate.

Components	Black oil (mol%)	Molecular weight	SG @60F/60F
METHANE	23.96	16.04	
ETHANE	3.98	30.07	
PROPANE	5.65	44.10	
ISOBUTANE	2	58.12	
N-BUTANE	4.12	58.12	
ISOPENTANE	2.09	72.15	
N-PENTANE	2.89	72.15	
C6	4.1	85	0.666
C7	6.25	94	0.714
C8	6.55	105	0.738
C9	5.2	119	0.761
C10	4.06	133	0.776
C11	3.31	147	0.790

C12	2.64	162	0.801
C13	2.36	173	0.815
C14	2.18	188	0.831
C15	1.95	203	0.837
C16	1.61	220	0.844
C17	1.28	233	0.849
C18	1.27	248	0.856
C19	1.05	262	0.863
C20+	11.5	483	0.947

Table 4 The compositional analysis of the natural gas and gas condensate.

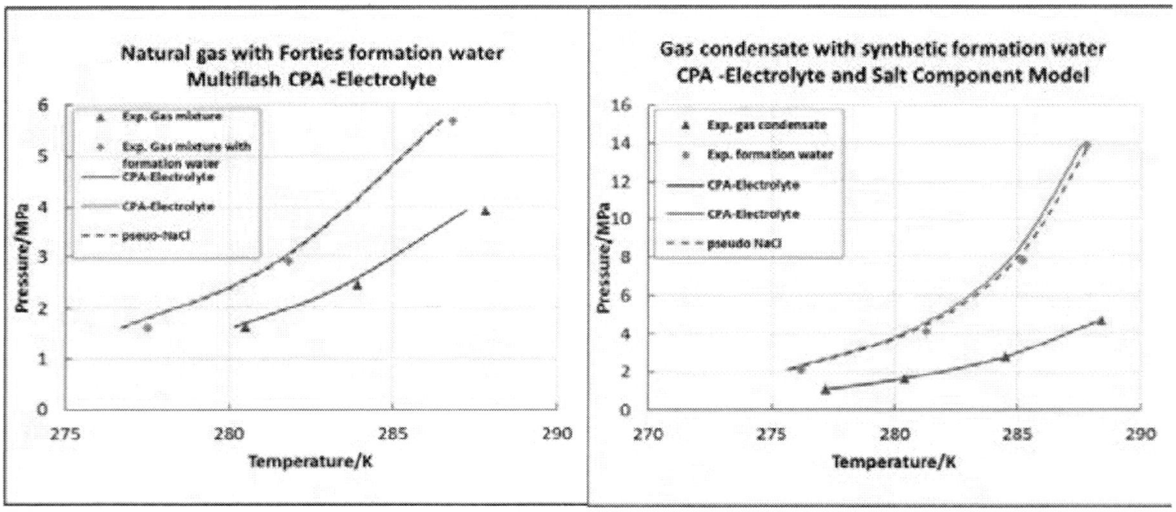

Figure 16 Hydrate dissociation curves of the natural gas with FFW and gas condensate with SFW, predicted by Multiflash™.

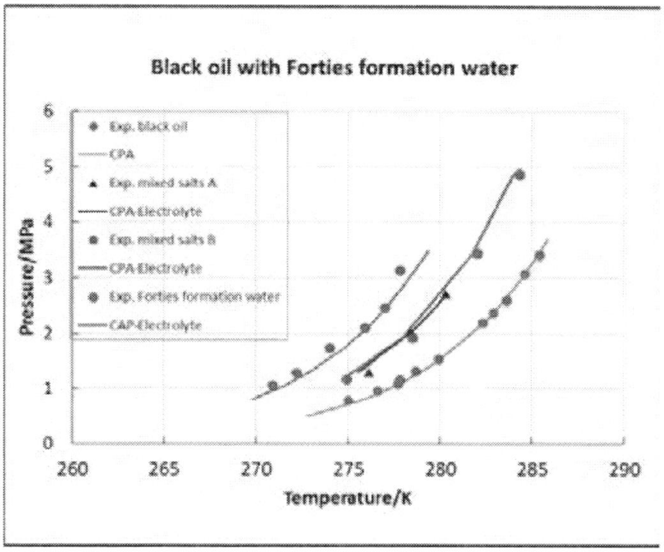

Figure 17 Hydrate dissociation curves of the black oil with mixed electrolytes and FFW, predicted by Multiflash™.

4 Conclusion

Two well-known thermodynamic models; CPA (Cubic-Plus-Association) with Electrolyte and the advanced version of RKS with modified NRTL type mixing rule were applied to study the suppression effect of single and mixed salts on hydrate formation. The fluid systems used in the study include the pure methane, binary methane-CO_2 gas mixture, natural gas, gas condensate and black oil, and the total salt concentration ranges from 3wt% to 21wt% including Forties formation water and synthetic formation water. For methane-CO_2 gas mixture, the mixed solvents with methanol concentrations of 5wt%, 10wt% and 15wt% are also considered to investigate the effect of mixed solvents on hydrate inhibition.

The study shows that both models are capable of giving accurate predictions on the suppression effect of single and mixed electrolytes on hydrate formation. The average accuracy of the predicted hydrate dissociation temperature is within 0.5 degree for CPA-Electrolyte, and about 1 degree for RKSA-Infochem with pseudo-NaCl salt approach. For the pseudo-NaCl approach, the salts and mixed electrolytes were approximately represented by a single pseudo-NaCl salt component, which works well for the cases with total salt concentration less than 20wt%. However this approach needs to be improved when dealing with mixed solvents involving salts with methanol or glycols and cases with high salt concentrations.

It is very interesting to know that the salts NaCl and $CaCl_2$ by weight are actually stronger than methanol as shown in Figure 13 and Figure 14. This demonstrates that taking into account of the supressing effect of salts in formation and produced water and brines on hydrate formation is important in deciding the correct hydrate inhibitor injection rates in oil and gas production.

In general the CPA-Electrolyte, as a rigorous thermodynamic model, is the recommended model for modelling effects of salts on hydrate inhibition in Multiflash™. However the RKSA-Infochem with a single pseudo-NaCl salt component has some advantages; simple, robust and is also able to give accurate predictions on the behaviour of fluid and solid phases, having the potential to be used in transient pipeline simulators.

5 References

1. E. Dendy Sloan and Carolyn A. Kon, "*Clathrate Hydrates of Natural Gases*", Third Edition, CRC Press Taylor & Francis Group, 2008.
2. Richard Szczepanski, Beryl Edmonds , Nigel Brown, Terry Hamilton, Technology 1998, 0il & Gas Journal March 9th 1998.
3. B. Edmonds, R. A. S. Moorwood and R. Szczepanski, presented at Deepwater and Ultra Deepwater drilling, London, 17-18 May, 1999.
4. B. Edmonds, R. A. S. Moorwood and R. Szczepanski, Paper presented at SMI Offshore Production Technology Conference, London, 31 January – 1st February, 2001.
5. P.F.Pickering, B. Edmonds, R.A.S.Moorwood, R. Szczepanski, and M.J. Watson, IIR Conference, Aberdeen 16th – 20th September 2001.
6. E.M Hendriks, B. Edmonds, R.A.S. Moorwood and R. Szczepanski, Fluid Phase Equilibria, 117, 193-200, 1996.
7. B. Edmonds, R. A. S. Moonwood, and R. Szczepanski, SPE 35569, 1996.
8. Ripmeester, J. A. and C. I. Ratcliffe, J. Phys. Chem., 94, p8773, 1990.
9. Tohidi, B, A. .Danesh, R.W. Burgass, A.C. Todd, SPE35568, 1990.
10. Ripmeester, J.S. Tse, C.J. Ratcliffe, G.E. Mclaurin, 35, 325 (1987)
11. J.H. van der Waals and J.C. Platteeuw, Adv. Chem. Phys., 2, 1-57, 1959.
12. W.R. Parrish and J.M. Prausnitz, Ind. Eng. Chem. Proc. Des. Dev., 11: 26-35, 1972.
13. Infochem Computer Services Ltd, "*Modelling gas hydrate formation and inhibition for black oils*", Deepstar contract work, November 2002.
14. Georgios M. Kontogeorgis and Georgios K. Folas, *"Thermodynamic Models for Industrial Applications"*, WILEY, A John Wiley and Sons Ltd, Publication, 2010.
15. Peiming Wanga, Kenneth S. Pitzer and John M. Simonson, J. Phys. Chem. Ref. Data, Vol. 27, No. 5, 1998.
16. K. S. Pitzer, "*Activity Coefficients in Electrolyte Solutions*", 2nd Edition (Chemical Rubber, Boca Raton, FL, 1991.
17. R. A. Robirson and R.H. Stokes, "Electrolyte Solutions", 3rd. ed., Butterworth, London, 1970.
18. Hooman Haghighi, Antonin Chapoy, and Bahman Tohidi, Ind. Eng. Chem. Res. 47, 3983–3989, 2008.
19. P. D. Dholabhai, P. Englezos, N. Kalogerakis and P. R. Bishnoi, The Canadian Journal of Chemical Engineering, Vol. 69, June 1991.
20. Pankaj D. Dholabhai and P. Raj Bishnoi, J. Chem. Eng. Data, 39, 191-194, 1994.
21. Pankaj D. Dholabhai, J. Scott Parent, P. Raj Bishnoi, Fluid Phase Equilibria 141, 235-246, 1997.
22. Bahman Tohidi, Ali danesh, R.W. Burgass and A.C. Todd, SPE 28884, 1994.
23. Kasper K., Ostergaard, Bahman Tohidi, Rod W. Burgass, Ali Danesh, and Adrian C. Todd, AIChE Journal, vol. 47, No. 7, July 2001.
24. B. Tohidi, A. Danesh, A. C. Todd and R. W. Burgass, Chemical Engineering Science, Vol. 52, No. 19, pp. 3257 3263, 1997.
25. Multiflash for Windows 6.1, Users' Guide, 2017.

CFD Simulation of Various Flow Regimes in Two-Phase Gas-Liquid Flows

Sanjeeb Pal[1]

Getnet Agegnehu[1], David Messersmith[2], Joel Peltier[1], Kelly Knight[1], Matthew Voegele[3]

[1] Advanced Simulation & Analysis, Bechtel Nuclear, Security & Environmental, Inc., Reston, VA
[2] LNG Technology, Bechtel Oil, Gas & Chemicals, Inc., Houston, TX
[3] Mechanical Engineering, Bechtel Oil, Gas & Chemicals, Inc., Houston, TX

Prepared for Presentation at the American Institute of Chemical Engineers, 2017 Spring Meeting and 13th Global Congress on Process Safety, San Antonio, Texas, March 26-30

Keywords: Gas liquid flows, two phase flow simulation, computational fluid dynamics (CFD), Large Scale Interface (LSI) model

Abstract

Many industrial flow applications involve two-phase gas-liquid flows in complex geometry. Computational fluid dynamics (CFD) simulations of two-phase flow systems by resolving the flow at the gas-liquid interface can successfully handle problems, where the interface length scales can be resolved by the computational grid. The Eulerian slip-formulation, which allows each phase to be represented by its own phase field (momentum, energy, etc.), can be used for accurate predictions where one of the phases is distributed as small size droplets or bubbles. This is an advantage when the length scale of the droplets/ bubbles is smaller than the length scales that can be represented on the mesh. This paper presents an application of a modified Eulerian slip-formulation method which adapts the inter-phase exchange (such as momentum exchange) to the local flow conditions. This method handles the large scale separate regions of gas and liquid using the Large Scale Interface (LSI) modeling approach. The interphase exchange modeling in dispersed flow regions can be sensitized to the local flow characteristics and mesh resolution, by specifying a user defined interface length scale that is used in calculating the interphase exchange terms. The application presented in this paper is a complex piping system with multiple out-of-plane bends connecting horizontal and vertical sections with changes in cross-sectional area. In the application presented, some flow regions may see large resolvable gas or liquid pockets, and in other regions the gas or liquid is present in bubble or droplet form that cannot be resolved by the mesh. The various flow regimes simulated are stratified-wavy, annular, and churn flow regimes for two-phase, immiscible gas-liquid flow in horizontal and vertical pipes with adiabatic walls.

Nomenclature

$\bar{u}_k$	velocity field for phase k
α_k	volume fraction of phase k
α_L	upper volume fraction limit for secondary phase corresponding to end of the 1st regime
α_U	lower volume fraction limit for secondary phase corresponding to start of the 2nd regime
ρ_k	density field of phase k
p	pressure field common to all phases
$\bar{g}$	body force field per unit mass
$\bar{f}_{sk}$	total surface tension force per unit volume from other phases on phase k
$\bar{\bar{\tau}}_k^M$	molecular shear stress tensor for phase k

$\bar{\bar{\tau}}_k^T$	turbulent shear stress tensor for phase k
$\bar{S}_{jk}^{Mom}$	interfacial force exerted by phase j on phase k per unit volume
S_{jk}^{Mass}	interfacial mass transfer from phase j to phase k per unit volume
$\bar{u}_{jk}$	slip velocity vector between phase j and phase k
A_{jk}	linearized drag coefficient for drag force on phase k from phase j
a_{jk}^s	surface area per unit volume for momentum exchange between phase k and phase j
τ_R	interface relaxation time scale at the Large Scale Interface
C_D^{bubble}	drag coefficient representing total fluid dynamic force on a bubble
$C_D^{droplet}$	drag coefficient representing total fluid dynamic force on a droplet
$W_{(\ )}$	weighting factors for different flow regimes used to calculate momentum source term
$(\ )_P$	denotes primary phase
$(\ )_S$	denotes secondary phase
$(\ )_{dispersed}$	denotes dispersed phase

1. Introduction

Simulation of two-phase gas-liquid flows in complex industrial piping systems requires simulation of various regimes of two phase flows. This paper presents an internal flow application for a modified Eulerian slip-formulation method which adapts the inter-phase exchange (such as momentum exchange) to the local flow conditions. This method handles the large scale separate regions of gas and liquid using the Large Scale Interface (LSI) modeling approach [1]. The interphase exchange modeling in dispersed flow regions can be sensitized to the local flow characteristics and mesh resolution, by specifying a user defined interface length scale that is used in calculating the interphase exchange terms. The Phenomena Identification and Ranking Table (PIRT) shown in Table 1 ranks the various phenomena based on their relative importance.

Table 1 Phenomena Identification and Ranking Table (PIRT) for two-phase gas-liquid internal flows.

Phenomena	Importance
Bubble drag for bubbles in liquid	Major
Bubble size distribution evolution (including coalescence and break-up)	Major
Carrier fluid unsteady flow effects on bubble drag	Moderate
Carrier fluid turbulence effects on bubble motion	Moderate
Near wall effects on bubble motion	Moderate
Near wall gas film modeling	Minor - Moderate
Droplet drag for droplets in gas stream	Major
Droplet size distribution evolution (including coalescence and break-up)	Major
Carrier fluid unsteady flow effects on droplet drag	Moderate
Carrier fluid turbulence effects on droplet motion	Moderate
Near wall effects on droplet motion	Moderate
Near wall liquid film modeling	Moderate - Major
Interface momentum-exchange at large scale interfaces	Major
Surface-tension effects at gas-liquid interface	Minor - Moderate
Multiphase turbulence modeling	Major
Near wall turbulence modeling	Moderate
Turbulence damping near interface	Moderate
Note(s): Importance is labeled as major, moderate, minor or unknown	

2. Methodology

The multi-fluid model governing equations are conservations equations for mass and momentum for each phase, which in fully resolved (DNS) form are written as:

$$\frac{\partial(\alpha_k \rho_k)}{\partial t} + \overline{\nabla}\cdot(\alpha_k \rho_k \bar{u}_k) = \sum_{j=1}^{n} S_{jk}^{Mass}$$

$$\frac{\partial(\alpha_k \rho_k \bar{u}_k)}{\partial t} + \overline{\nabla}\cdot(\alpha_k \rho_k \bar{u}_k \otimes \bar{u}_k) = -\alpha_k \overline{\nabla} p + \overline{\nabla}\cdot\left(\alpha_k \bar{\bar{\tau}}_k^M\right) + \alpha_k \rho_k \bar{g} + \sum_{j=1}^{n} \bar{S}_{jk}^{Mom}$$

The slip velocity vector between phase j on phase k is given by, $\bar{u}_{jk} = \bar{u}_k - \bar{u}_j$. For a two fluid system, we have two sets of conservation equations for mass and momentum of each phase.

The interfacial mass and momentum exchange terms need to be modeled. The modeling of these terms requires the knowledge and further modeling of

a) Interfacial contact area per unit volume
b) Fluid property differences across the interface
c) Slip velocity between phases and turbulence levels associated with both phases

In the LSI model for a pair of fluids, the interfacial area is modeled based on the inferred flow regimes for a gas-liquid system. The regime classification used in this study is based on volume fractions in given grid cell as given below:

$0 \le \alpha_S \le \alpha_L$ The secondary phase is treated as the dispersed phase (1st regime)
$\alpha_L \le \alpha_S \le \alpha_U$ The cell is deemed to contain a large scale interface (LSI regime)
$\alpha_U \le \alpha_S \le 1$ The primary phase is treated as the dispersed phase (2nd regime)

Here we take the gas phase as the primary phase and the liquid phase as the secondary phase. Indices j and k could represent the primary or secondary phase.

a) Regime 1 - Droplet flow with small amount of liquid present in the cell which is primarily gas

$$\bar{S}_{jk}^{Mom} = A_{jk}\bar{u}_{jk}$$

$$A_{jk} = \frac{1}{8} C_D^{droplet} a_{jk}^{s} \rho_j \left|\bar{u}_{jk}\right|$$

b) Regime 2 - Bubbly flow with small amount of gas phase present in the cell which is primarily liquid. The interface momentum exchange is expressed in linearized form as

$$\bar{S}_{jk}^{Mom} = A_{jk}\bar{u}_{jk}$$

$$A_{jk} = \frac{1}{8} C_D^{bubble} a_{jk}^{s} \rho_j \left|\bar{u}_{jk}\right|$$

c) LSI regime - Separated gas-liquid flow regime where gas and liquid volume fractions in a cell are comparable

$$\bar{S}_{jk}^{Mom} = A_{jk}\bar{u}_{jk} + \bar{f}_{sk}$$

$$A_{jk} = \frac{1}{\tau_R}\alpha_P(1-\alpha_P)\{\alpha_P \rho_P + (1-\alpha_P)\rho_S\}$$

The surface tension forces are additionally incorporated as given below.

$$\bar{f}_{sk} = \frac{\alpha_k \sigma \overline{\nabla} \alpha_k}{R}$$

The radius of curvature, R is calculated as

$$\frac{1}{R} = -\overline{\nabla} \cdot \hat{n} = -\overline{\nabla} \cdot \left(\frac{\overline{\nabla} \alpha_k}{|\overline{\nabla} \alpha_k|} \right)$$

As we move from regime to regime in the mesh the calculation of momentum source terms for each phase is calculated as weighted sum of evaluations based on the flow regimes discussed above. The weighting ensures smooth non-abrupt change in source term as the regime changes from one to the next regime.

$$A_{jk} = W_{droplet} A_{jk}^{droplet} + W_{LSI} A_{jk}^{LSI} + W_{bubble} A_{jk}^{bubble}$$

$$W_{droplet} = \frac{1}{1 + e^{C(\alpha_S - \alpha_L)}}$$

$$W_{bubble} = \frac{1}{1 + e^{C(\alpha_U - \alpha_S)}}$$

$$W_{LSI} = 1 - \left(W_{droplet} + W_{bubble} \right)$$

A Predictive Capability Maturity Requirements (PCMR) matrix in Table 2 summarizes the expectations of physical sub-model fidelity for accurate simulation of the two-phase flow conditions using the above model.

Table 2 Predictive Capability Maturity Requirements (PCMR) table for sub-models.

Phenomena	Required model fidelity
Bubble drag estimate in various flow conditions	High
Length scale modeling for local bubble phase size distribution	High
Unsteady carrier fluid flow effects on bubble drag	Moderate
Turbulence effects on bubble drag	Moderate
Near wall modification of bubble drag	Moderate
Droplet drag estimate in various flow conditions	High
Length scale modeling for local droplet size distribution	High
Unsteady carrier fluid flow effects on droplet drag	Moderate
Turbulence effects on droplet drag	Moderate
Near wall modification of droplet drag	Moderate
Interface relaxation time scale modeling based on fluid property and local flow conditions	High
Multiphase turbulence modeling	High
Near wall turbulence modeling	Moderate
Turbulence damping near interface	Moderate
Note(s): Desired model fidelity is labeled as high, moderate, low, or none (i.e., model could neglect phenomena).	

Figure 1 shows the various length scales that are not represented or resolved, or resolved on a given computational grid.

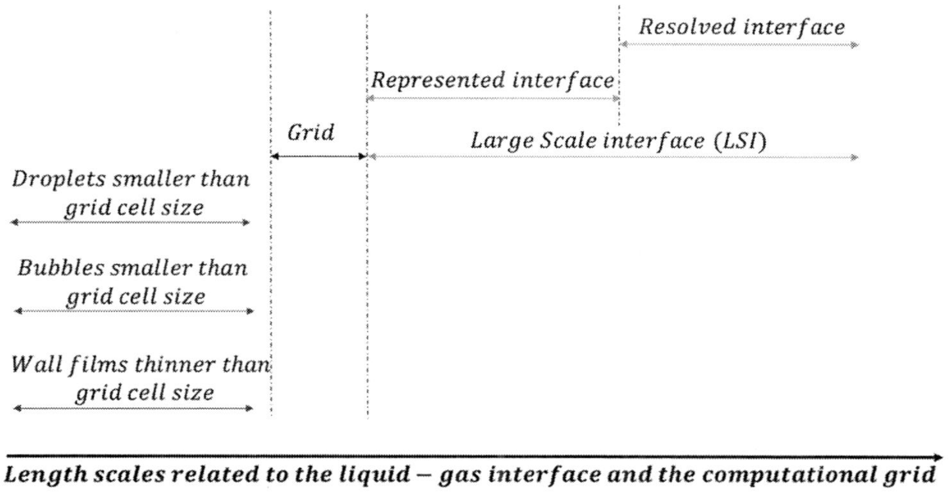

Figure 1 Range of length scales relative to the computational grid.

3. Modeling of Turbulence

The ensemble averaging of the multiphase governing equations introduces additional terms that are modelled. The phase fraction, momentum and pressure fields in the DNS form equations are replaced by ensemble averages, and the Reynolds stress tensor is introduced for each phase which is added to the phase molecular shear stress tensor as shown below.

$$\overline{\nabla} \cdot \left(\alpha_k \overline{\overline{\tau}}_k^M\right) \text{ is replaced by } \overline{\nabla} \cdot \left\{\alpha_k\left(\overline{\overline{\tau}}_k^M + \overline{\overline{\tau}}_k^T\right)\right\}$$

The turbulence modeling equations are solved separately for each phase. Additional damping is introduced near the LSI interface to account for turbulence damping near the liquid-gas interface.

4. CFD Model Geometry and Flow Conditions

The application presented in this paper is a complex piping system with multiple out-of-plane bends connecting horizontal and vertical sections with changes in cross-sectional area. In the application presented, some flow regions may see large resolvable gas or liquid pockets, and in other regions the gas or liquid is present in bubble or droplet form that cannot be resolved by the mesh. The various flow regimes simulated are stratified-wavy, annular, and churn flow regimes for two-phase, immiscible gas-liquid flow in horizontal and vertical pipes with adiabatic walls. The flow rates are summarized in Table 3. The gas to liquid kinematic viscosity ratio is close to 2, smaller compared to about 16 for air-water mixtures at room temperatures.

Table 3 Summary of flow rates and fluid properties used in the analysis.

	Liquid flow rate	Gas flow rate	Liquid density	Gas density	Liquid volume fraction	Gas volume fraction	Liquid viscosity	Gas viscosity	Liquid-Gas surface tension
	[kg/s]	[kg/s]	[kg/m^3]	[kg/m^3]	[ ]	[ ]	[Pa-s]	[Pa-s]	[Nm^{-1}]
100 % flow	11.74	7.33	538.02	30.85	0.0841	0.9159	1.2596E-4	1.1451E-5	0.7783E-2
35 % flow	4.11	2.56	538.02	30.85	0.0841	0.9159	1.2596E-4	1.1451E-5	0.7783E-2

Figure 2 and Figure 3 show the details of the baseline design #1 piping system from heat exchanger outlet to the inside of the debutanizer column. Two additional designs, alternate design #2 and alternate design #3 are also evaluated using CFD modeling.

Table 4 summarizes the calculated superficial Reynolds numbers in various sections of the piping.

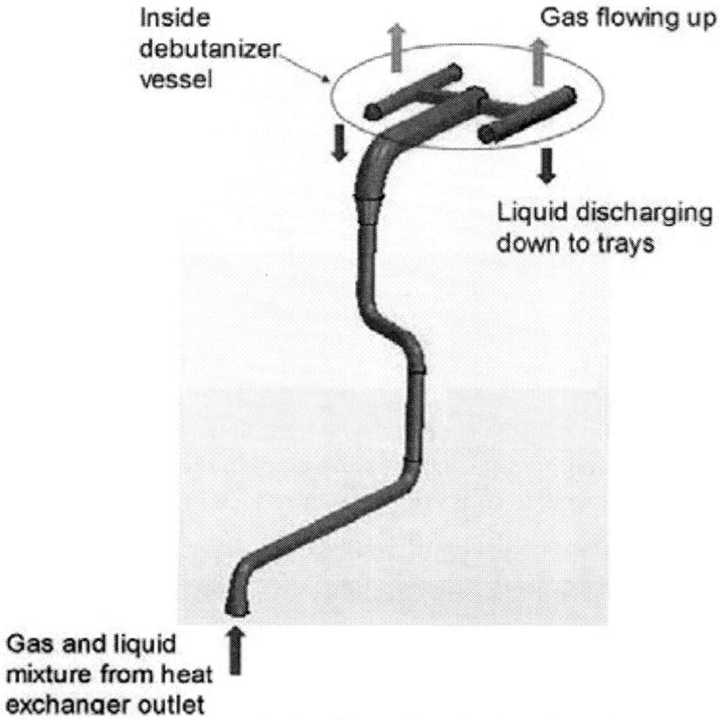

Figure 2 Details of the baseline piping design #1 analyzed using the Eulerian two-fluid LSI model.

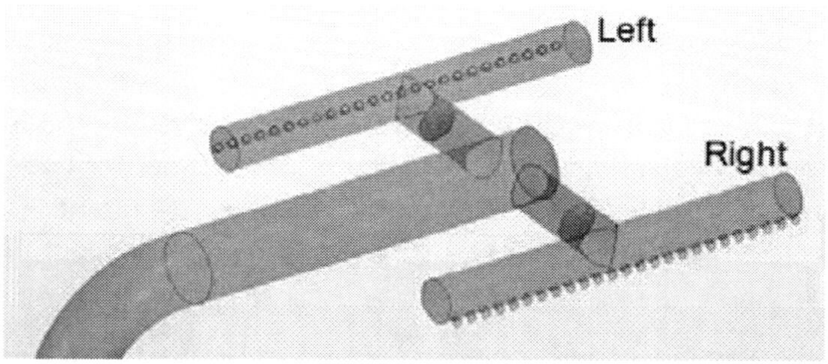

Figure 3 Details of the piping inside the confines of the debutanizer tower showing the distributor and series of discharge openings on the left and right side.

Table 4 Summary of superficial Reynolds numbers in various sections of the piping system.

	Liquid - 7.95 inch pipe	Liquid - 7.95 inch distributor branches	Liquid - 13.25 inch horizontal pipe	Gas - 7.95 inch pipe	Gas - 7.95 inch distributor branches	Gas - 13.25 inch pipe horizontal pipe
100% flow	5.87E+05	2.94E+05	3.53E+05	3.26E+06	1.63E+06	1.96E+06
35% flow	2.06E+05	1.03E+05	1.23E+05	1.14E+06	5.71E+05	6.86E+05

The flow regimes assessed in the horizontal and vertical sections of the pipe using flow regime maps such as Taitel-Dukler (1976) map [2] for horizontal pipe sections, and the Hewitt-Roberts (1969) map [3] for vertical section are summarized below in Table 5. Figure 4 and Figure 5 show the common regimes of two-phase gas-liquid flows in horizontal and vertical pipes without heat transfer.

Table 5 Summary of flow regime assessments for the piping system.

	100% flow rate			35% flow rate		
	Liquid flow rate [kg/s]	Gas flow rate [kg/s]	Inferred flow regime	Liquid flow rate [kg/s]	Gas flow rate [kg/s]	Inferred flow regime
Horizontal pipe (below diffuser) with 7.95 inch internal diameter	11.74	7.33	Annular-Dispersed Liquid	4.11	2.56	Stratified-Wavy (but close to annular)
Vertical pipe (below diffuser) with 7.95 inch internal diameter	11.74	7.33	Annular	4.11	2.56	Churn flow
Horizontal pipe (in distributor) with 13.25 inch internal diameter	11.74	7.33	Stratified-Wavy (but close to annular)	4.11	2.56	Stratified-Wavy
Horizontal pipes (T-junction left and right side in distributor) with 7.95 inch internal diameter	5.87	3.67	Annular (close to Stratified-Wavy)	2.06	1.28	Stratified-Wavy

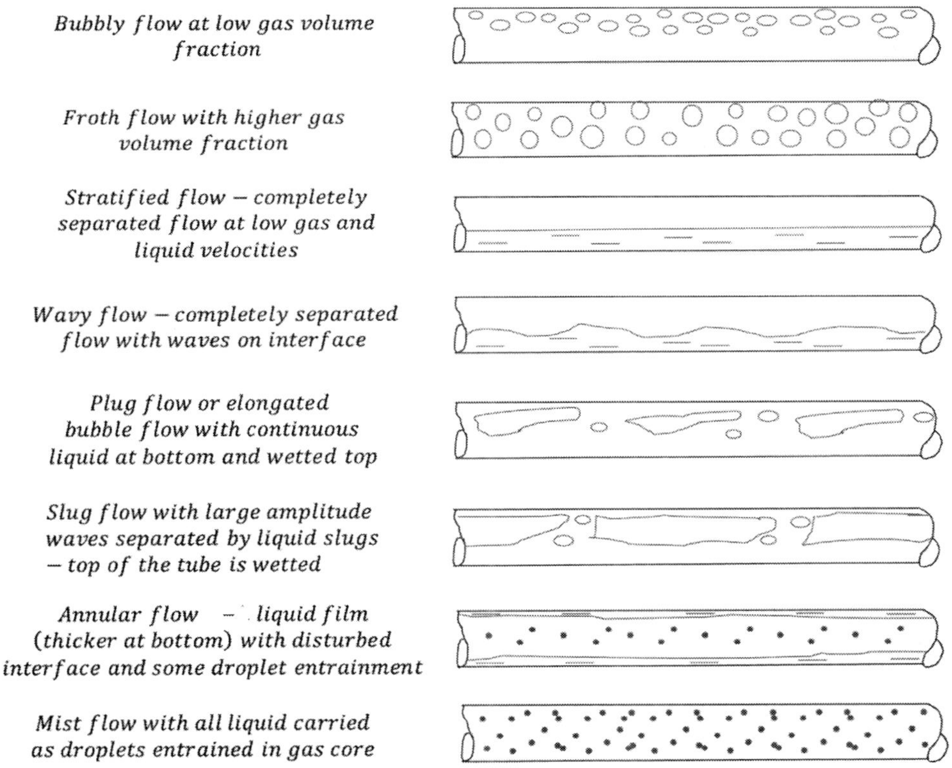

Figure 4 Two-phase flow regimes in horizontal pipes without heat transfer effects. Gas flow is increasing from top to bottom and liquid flow is increasing from bottom to top.

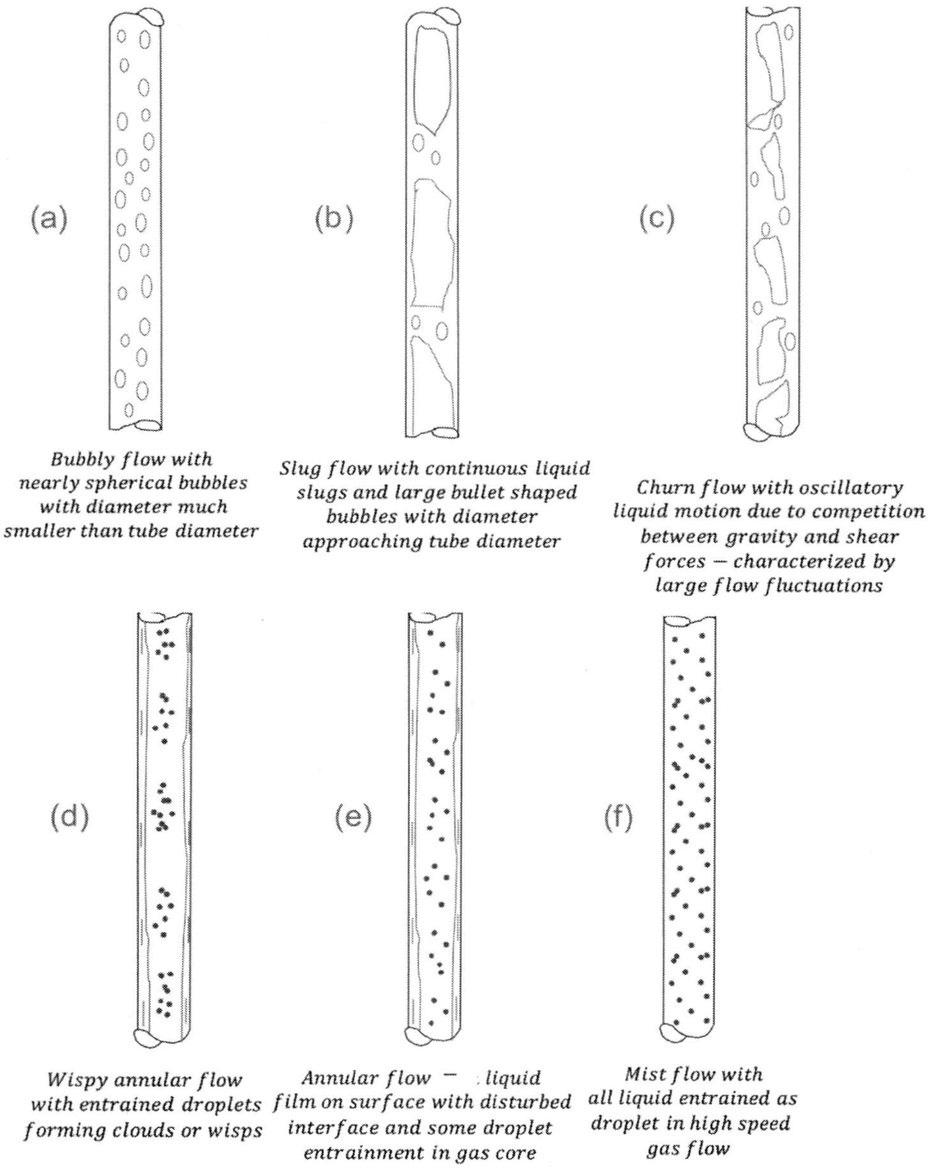

Figure 5 Two-phase flow regimes in vertical pipes without heat transfer. Gas flow is increasing from (a) to (f) and liquid flow is increasing from (f) to (a).

5. CFD Model Numerical and Physical Model Settings

The STAR-CCM+ CFD code [4] was used to simulate the two-phase flow. The solution uses a SIMPLE algorithm based segregated solver solving for the common pressure field, the liquid and gas momentum, volume fraction and turbulence fields.

The CFD model uses a polyhedral mesh with two prism layers on the wall surfaces to resolve the boundary layers. The size of the mesh is 804,475 cells. The mesh included "extruded" tubes at the left and right side distributor holes where the flow leaves the distributor. Similar mesh resolution was used for the three designs analyzed. A fixed time step is used for the solution and the number of inner iterations per time step is set to 15. The gas and liquid flow split results for the left and right side outlets on the distributor from the CFD model were checked to confirm that constant volume flow was maintained in the numerical solution at all times.

The same mesh was used for the 100% and 35% flow conditions. The same time step was used for the 100% and 35% flow simulations. A second order upwind convection scheme with Venkatakrishnan limiters is used for all transported variables other than turbulence quantities.

The $SST\ k - \omega$ model is used to model turbulence in each phase with the Durbin Limiter. An all y^+ near wall treatment is used to handle varying y^+ on the walls. The turbulence damping is used in near interface with the interface damping constant set to 3000. The turbulence viscosity ratio is clipped to a high level of 1E10 considering the possibility of steep waves suddenly breaking. The first order convection scheme is used for turbulence quantities in both phases.

The primary phase is set to gas and the secondary phase is set to liquid. The LSI regime is identified by the upper and lower limits of 0.3 and 0.7 for the secondary phase which mark the end of first regime and the start of second regime respectively. The LSI drag model is based on constant interface relaxation time scale, $\tau_R = 0.01\ s$.

The Schiller-Naumann drag law is used for both droplets and bubbles. Turbulence dispersion effects are modeled outside the LSI regime for bubbles and droplets. The interfacial area is modeled with prescribed length scales (diameter) given below and assume a spherical bubble or droplet shape.

$$L = \begin{Bmatrix} max\left(1E-5, 0.001 \cos\left[\frac{\pi}{2}\frac{0.1 - \alpha_{dispersed}}{0.1}\right]\right) m\ for\ \alpha_{dispersed} \leq 0.1 \\ 0.001\ m\ for\ \alpha_{dispersed} > 0.1 \end{Bmatrix}$$

The inflow boundary is located at the lower end of the piping where flow comes in from a heat exchanger. A homogenous mixed volume fraction and common uniform velocity normal to the inlet cross-section is specified for the gas and liquid phases. Turbulence quantities are also specified for the two phases at the inlet.

A prescribed static pressure is applied at the opens ends of the tubes "extruded" on the left and right hand side discharges in the distributor. Flow quantities are extrapolated for flow leaving the domain.

The whole domain is initialized to a homogenous mixture with the same phase volume fractions as at the inlet, and an initial pressure field equal to the hydrostatic pressure.

6. Results

Figure 6 shows the typical liquid volume fraction distribution (in top view) in the three piping system designs at specific instants for the 100% flow rate case. The presence of stratified-wavy flow condition in the 13.25 inch distributor horizontal section is evident. The expected annular-dispersed flow regime in the 7.95 inch piping leading to the distributor is not present in the CFD simulation result. Instead, a thick film (swirling) is seen in the horizontal piping at the bottom with the top of the pipe relatively devoid of liquid. This is believed to be due to the effect of the upstream bends. Figure 7 through Figure 9 show the side to side gas and liquid flow splits coming out of the distributor (at the extruded tubes) for the three designs at the 100% flow condition. The alternate design #3 has a high level of imbalance in the time averaged liquid flows through the left and right sides. Figure 10 and Figure 11 show the liquid motion using the 0.5 liquid volume fraction isosurface colored by the liquid velocity magnitude.

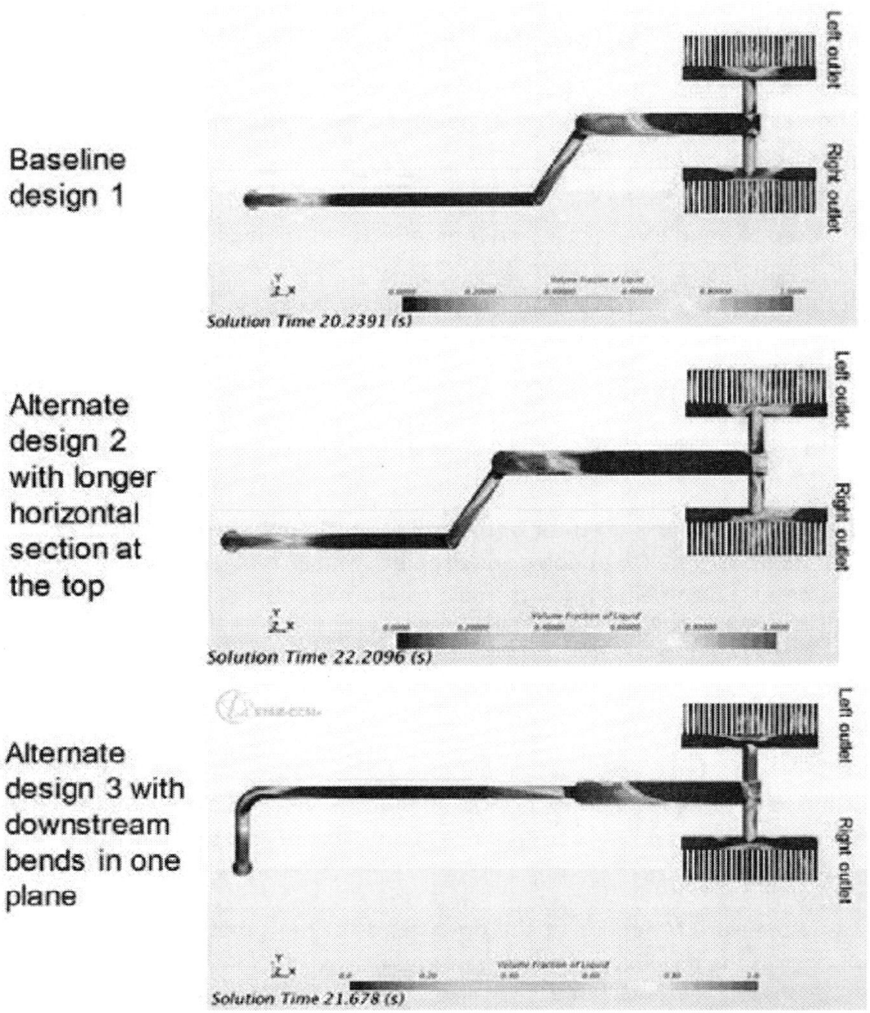

Figure 6 Liquid distribution (volume fraction contours) at specific times for 100% flow condition in the three piping system designs.

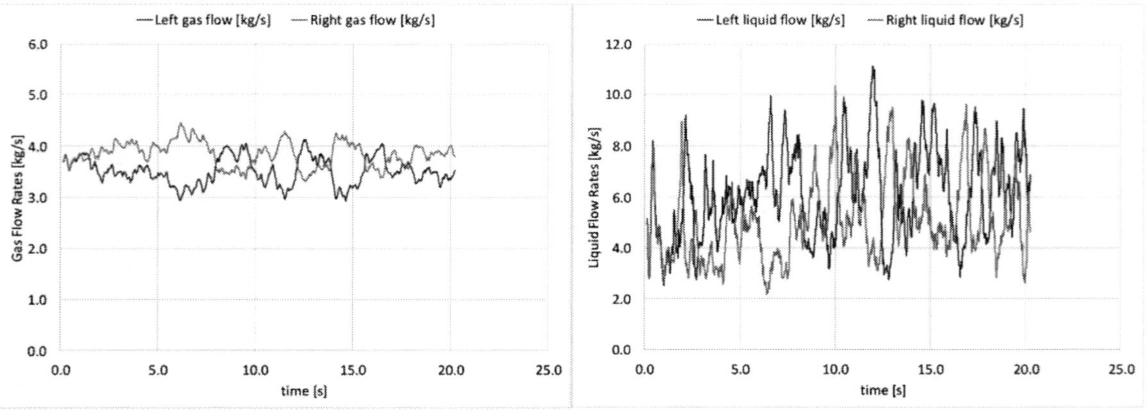

Figure 7 Side to side gas and liquid flow distribution versus time for the 100% flow conditions in the baseline design #1.

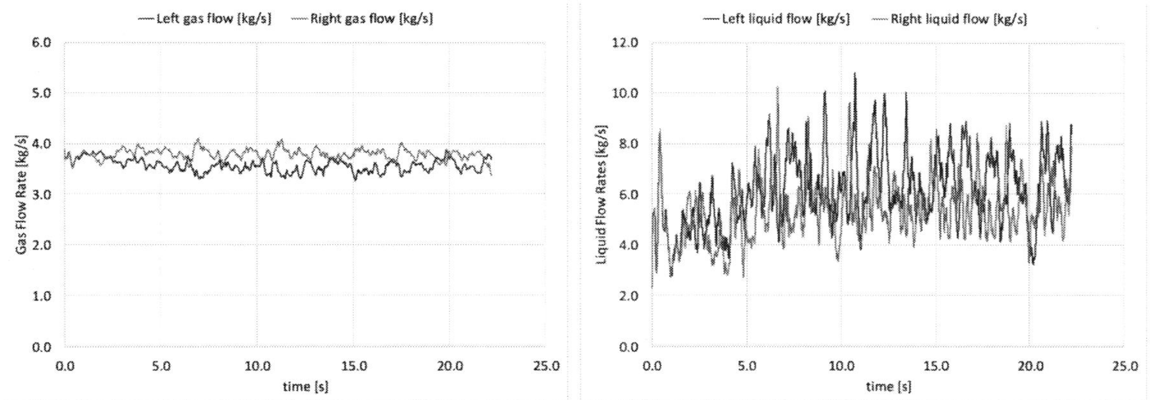

Figure 8 Side to side gas and liquid flow distribution versus time for the 100% flow conditions in the alternate design #2.

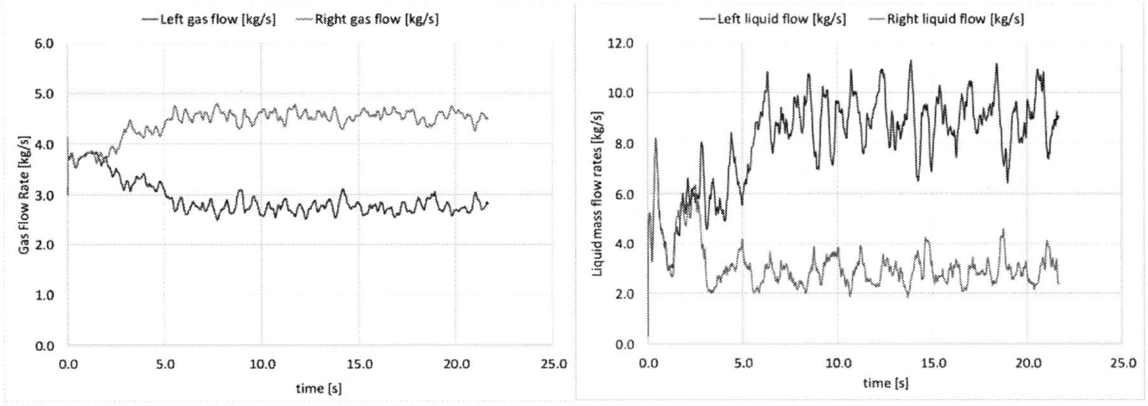

Figure 9 Side to side gas and liquid flow distribution versus time for the 100% flow conditions in the alternate design #3.

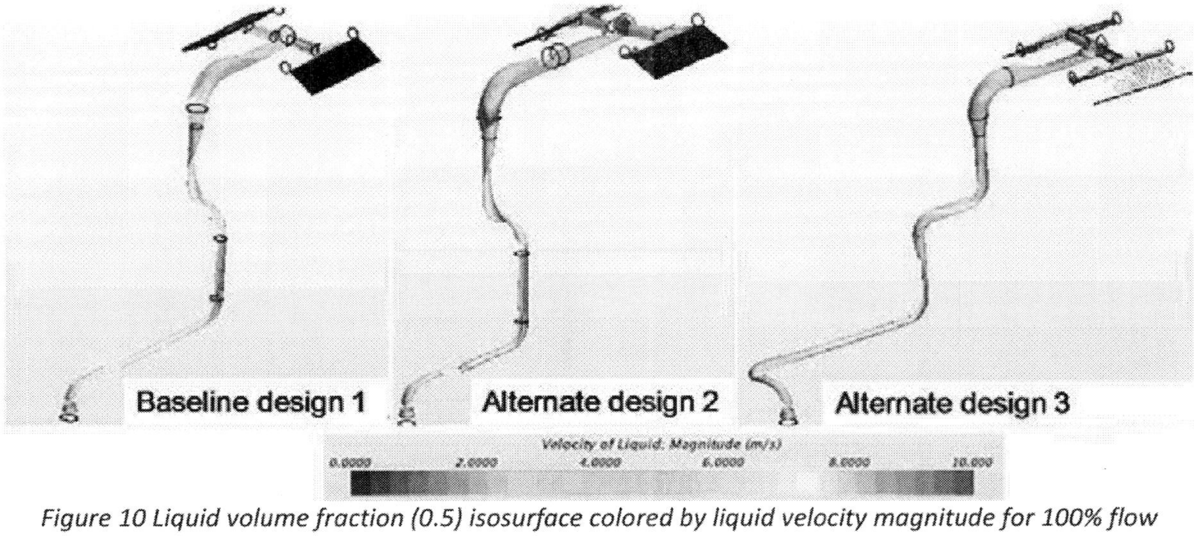

Figure 10 Liquid volume fraction (0.5) isosurface colored by liquid velocity magnitude for 100% flow conditions.

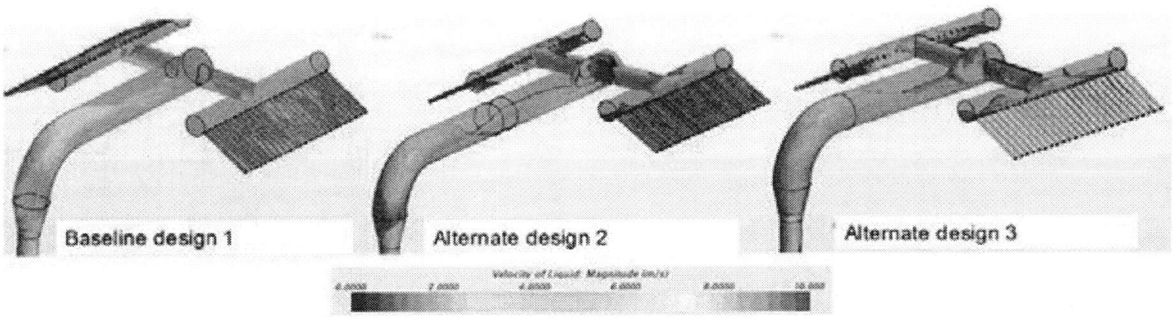

Figure 11 Liquid volume fraction (0.5) isosurface in the distributor region colored by liquid velocity magnitude at 100% flow conditions.

Figure 12 through Figure 14 show the side to side gas and liquid flow splits coming out of the distributor (at the extruded tubes) for the three designs at the 35% flow conditions. The time averaged liquid flows through the left and right sides are nearly equal for all the designs. Figure 15 and Figure 16 show the liquid motion using the 0.5 liquid volume fraction isosurface colored by the liquid velocity magnitude.

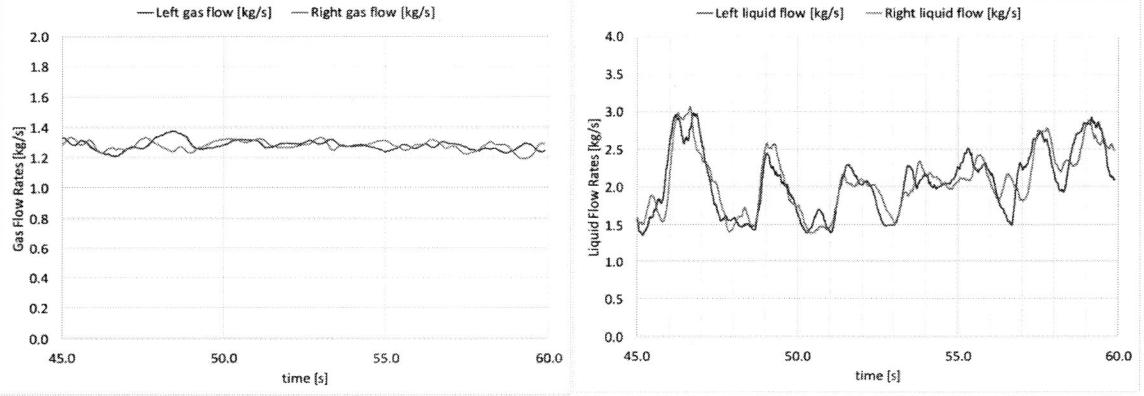

Figure 12 Side to side gas and liquid flow distribution versus time for 35% flow conditions in the baseline design #1.

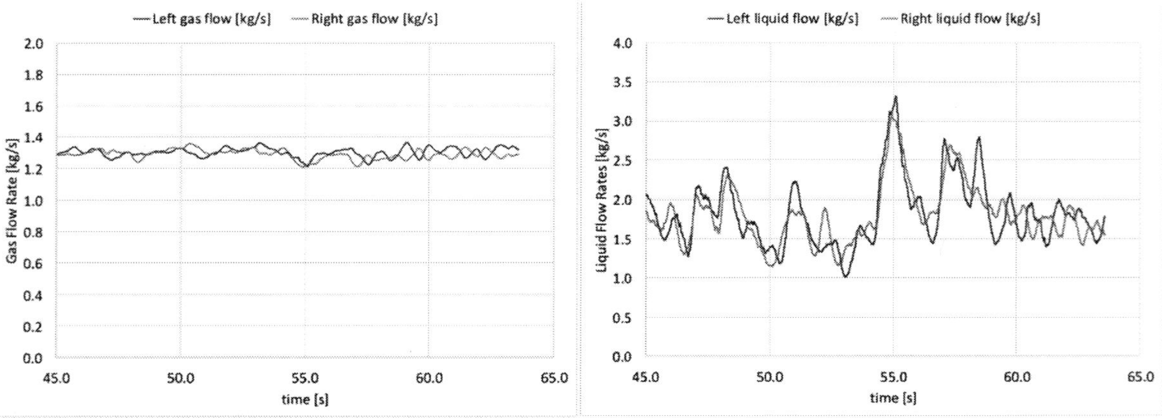

Figure 13 Side to side gas and liquid flow distribution versus time for 35% flow conditions in the alternate design #2.

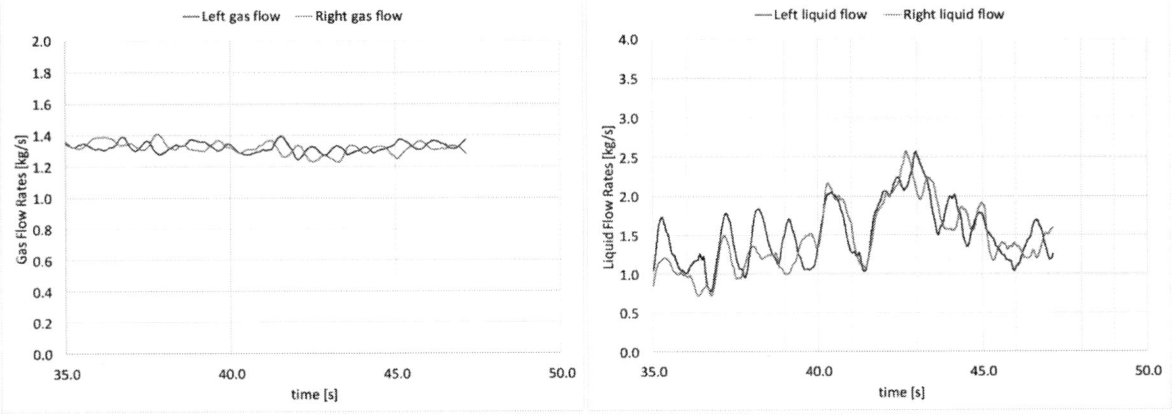

Figure 14 Side to side gas and liquid flow distribution versus time for 35% flow conditions in the alternate design #3.

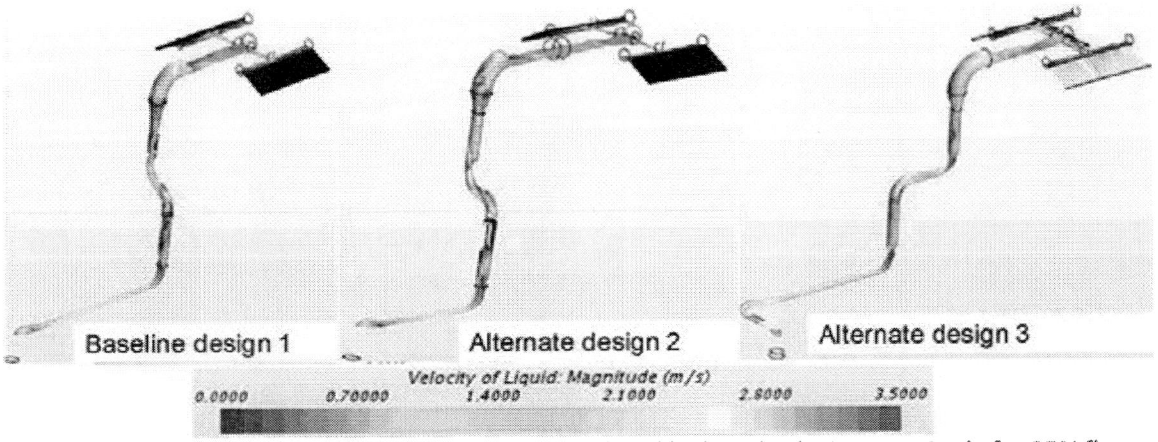

Figure 15 Liquid volume fraction (0.5) isosurface colored by liquid velocity magnitude for 35% flow conditions.

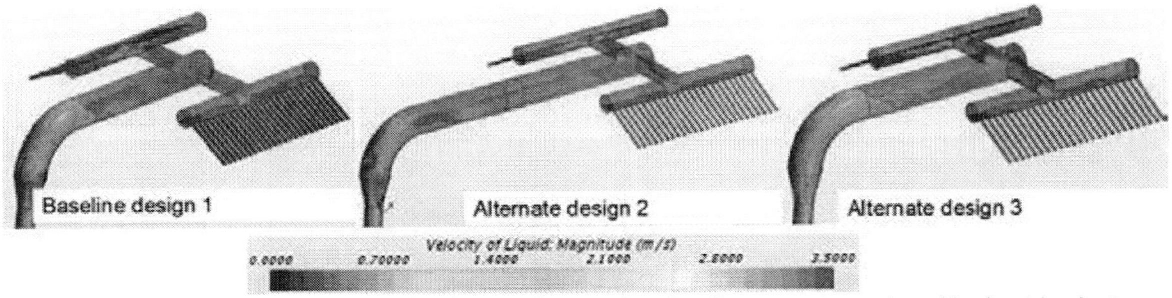

Figure 16 Liquid volume fraction (0.5) isosurface in the distributor region colored by liquid velocity magnitude for 35% flow conditions.

7. Model Parameter Sensitivity Study

The alternate design #3 CFD model is used to carry out a parameter sensitivity study on the interface relaxation time scale used to model the large scale interfacial drag. This model parameter was changed to half (case "2x") and one-fourth (case "4x") of the default value of 0.01 s for the sensitivity study. This parameter was changed in the baseline design #1 CFD model at a specific instant, and the three models with different values of this parameter were run further with the same time step size. Liquid flow rates were collected at different sections shown in Figure 17 after some time was allowed for the flows to adjust to the new parameters. The study addressed both the 100% and 35% flow cases. The liquid flow rate time history at the various sections in Figure 17 is compared for three different interface relaxation time scale values for the 100% and 35% flow conditions in Figure 18 through Figure 28. The level of temporal fluctuation in section liquid flow rates is quantified as RMS value about the time average in Table 6. The time averaged side to side liquid and gas flow splits based on section 10 and 11 liquid flow rate monitoring is summarized in Table 7.

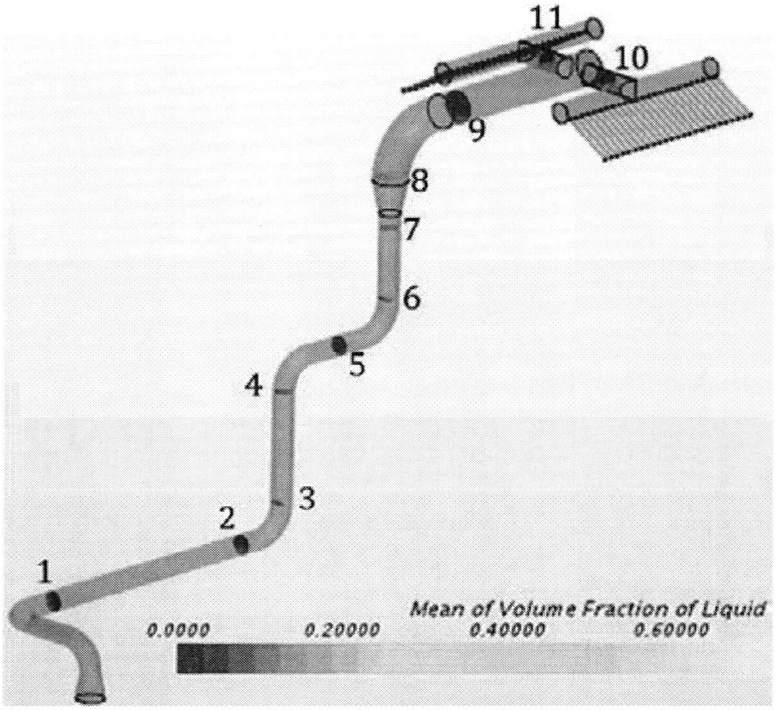

Figure 17 Location of cut sections used to monitor liquid flow rates for the alternate design #3 CFD model sensitivity study on the interface relaxation time scale.

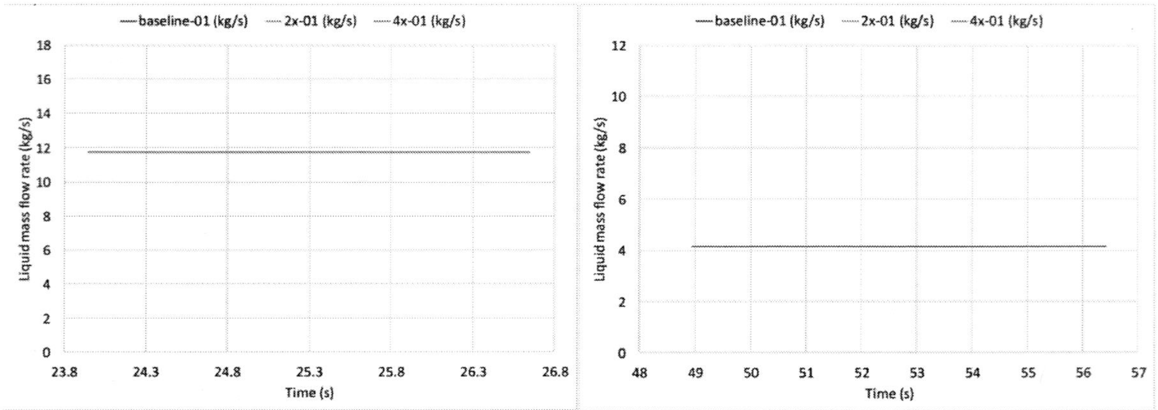

Figure 18 Liquid flow rate comparison at section 1 for three different values of interface relaxation time scale for 100% (left) and 35% (right) flow conditions in alternate design #3.

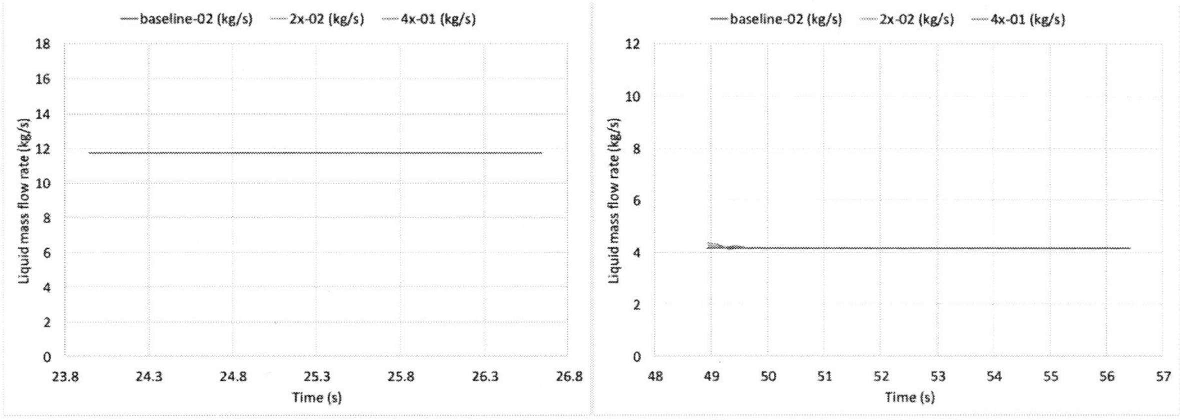

Figure 19 Liquid flow rate comparison at section 2 for three different values of interface relaxation time scale for 100% (left) and 35% (right) flow conditions in alternate design #3.

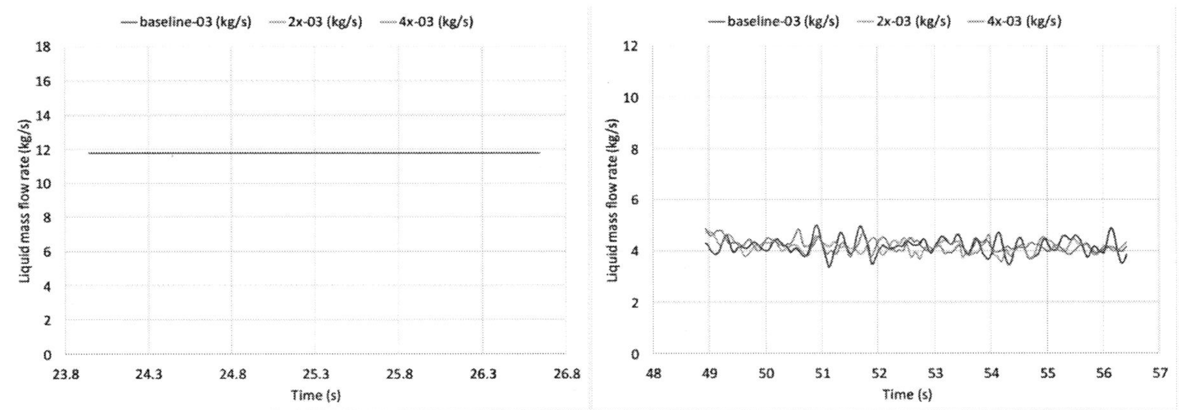

Figure 20 Liquid flow rate comparison at section 3 for three different values of interface relaxation time scale for 100% (left) and 35% (right) flow conditions in alternate design #3.

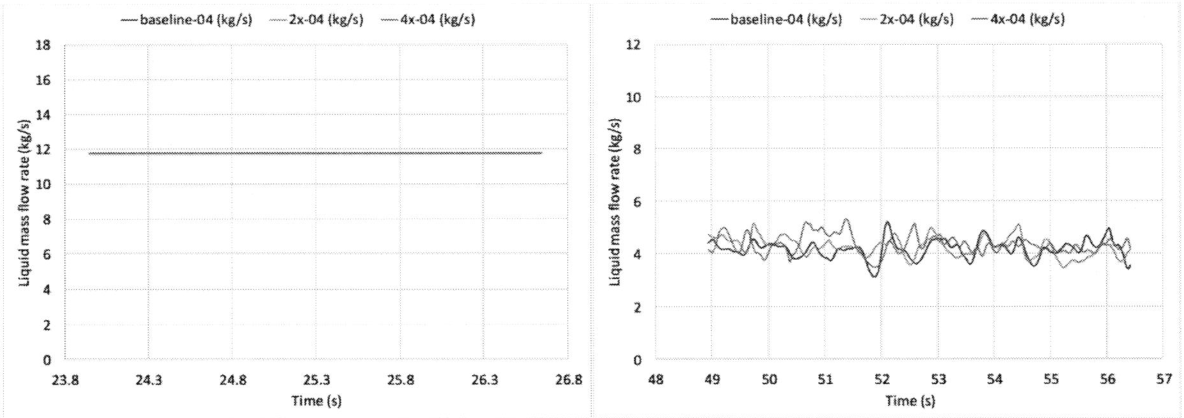

Figure 21 Liquid flow rate comparison at section 4 for three different values of interface relaxation time scale for 100% (left) and 35% (right) flow conditions in alternate design #3.

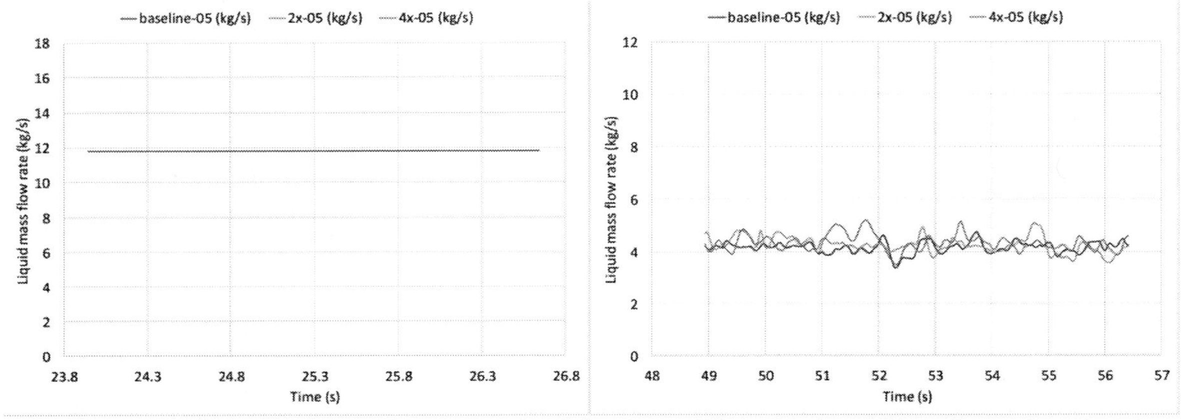

Figure 22 Liquid flow rate comparison at section 5 for three different values of interface relaxation time scale for 100% (left) and 35% (right) flow conditions in alternate design #3.

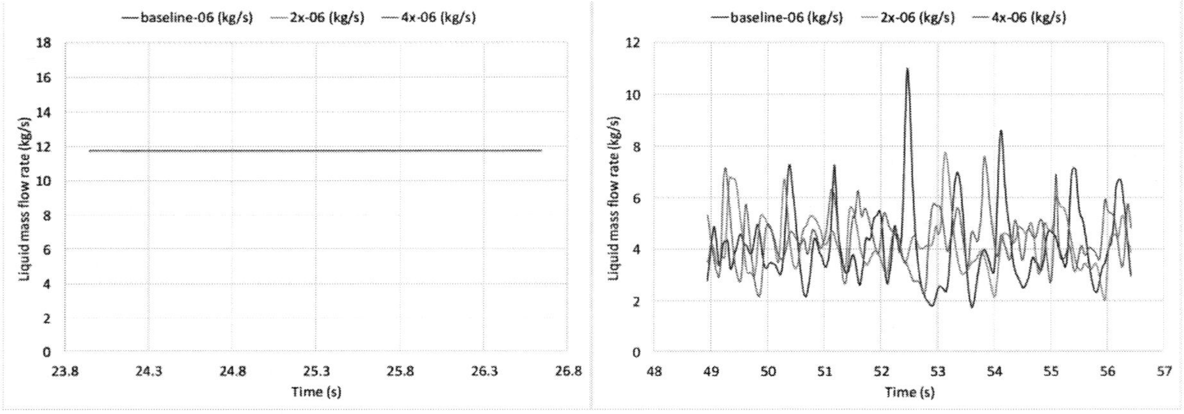

Figure 23 Liquid flow rate comparison at section 6 for three different values of interface relaxation time scale for 100% (left) and 35% (right) flow conditions in alternate design #3.

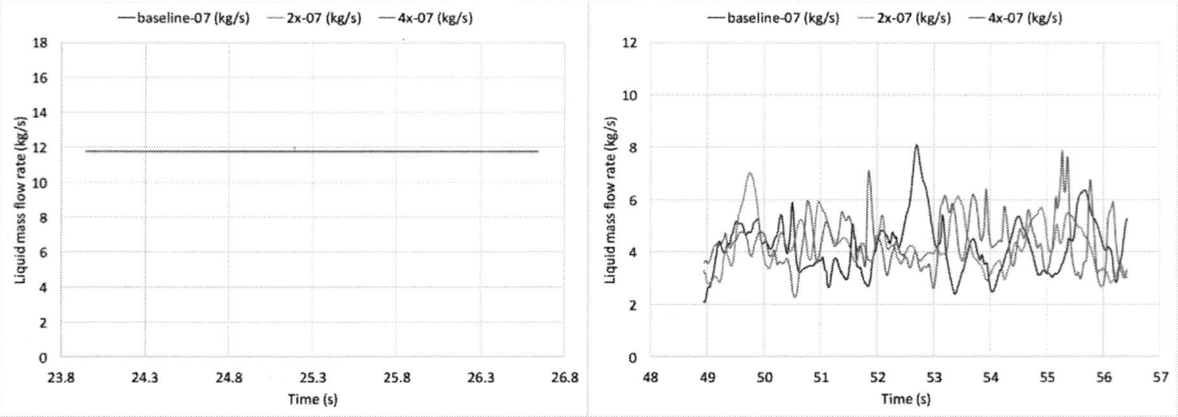

Figure 24 Liquid flow rate comparison at section 7 for three different values of interface relaxation time scale for 100% (left) and 35% (right) flow conditions in alternate design #3.

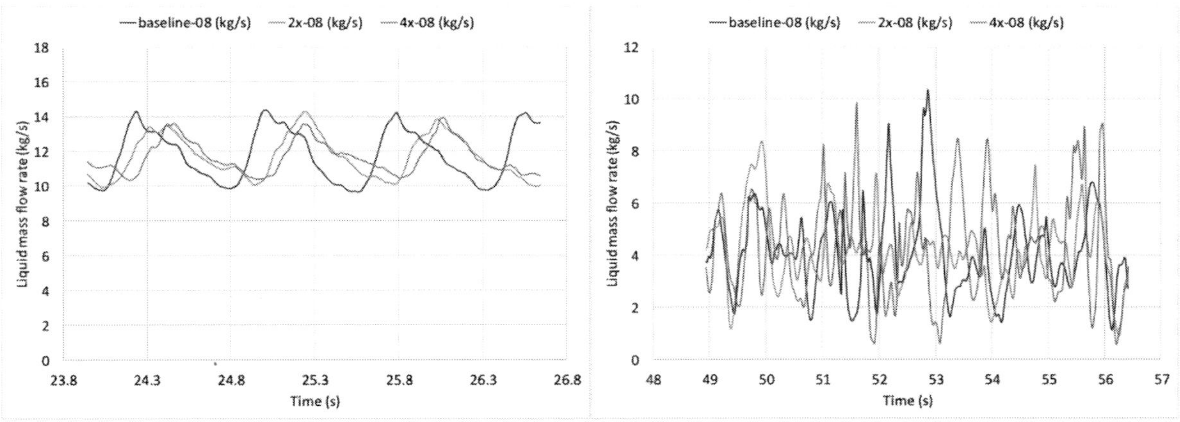

Figure 25 Liquid flow rate comparison at section 8 for three different values of interface relaxation time scale for 100% (left) and 35% (right) flow conditions in alternate design #3.

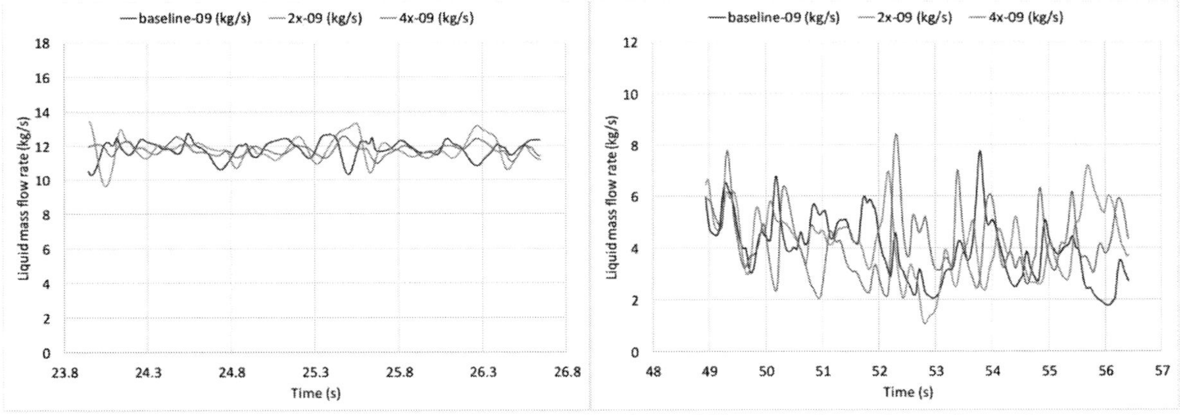

Figure 26 Liquid flow rate comparison at section 9 for three different values of interface relaxation time scale for 100% (left) and 35% (right) flow conditions in alternate design #3.

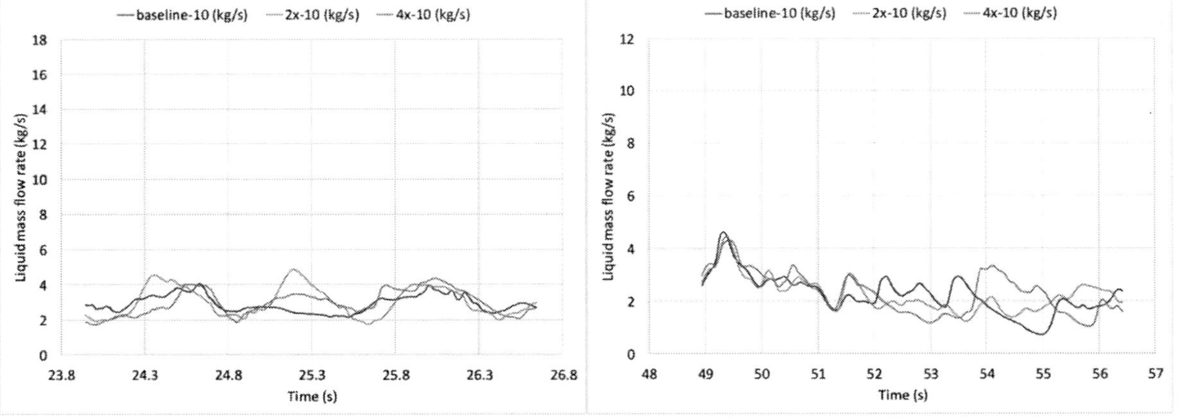

Figure 27 Liquid flow rate comparison at section 10 for three different values of interface relaxation time scale for 100% (left) and 35% (right) flow conditions in alternate design #3.

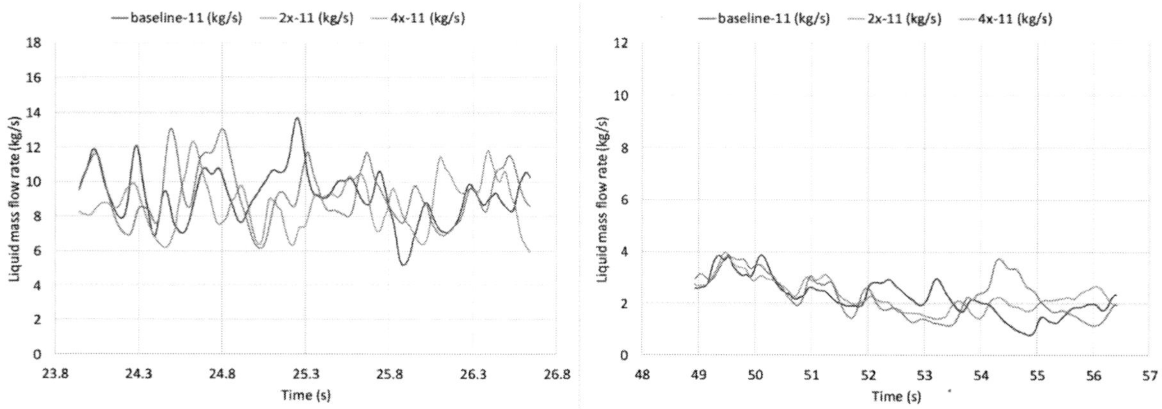

Figure 28 Liquid flow rate comparison at section 11 for three different values of interface relaxation time scale for 100% (left) and 35% (right) flow conditions in alternate design #3.

Table 6 RMS fluctuation level (temporal) for section liquid flow rates expressed as % of mean (time average) liquid flow rate.

Section	100% full flow rate			35% turndown flow rate		
	baseline	Case 2x	Case 4x	baseline	Case 2x	Case 4x
01	0.0001%	0.0001%	0.0001%	0.0010%	0.0267%	0.0434%
02	0.0019%	0.0003%	0.0014%	0.0010%	0.4127%	0.7464%
03	0.0033%	0.0006%	0.0009%	7.5835%	5.7533%	5.3232%
04	0.0076%	0.0023%	0.0033%	8.2627%	8.4588%	8.2064%
05	0.0081%	0.0021%	0.0057%	5.2532%	7.2146%	7.2805%
06	0.0076%	0.0056%	0.0148%	36.5867%	23.1021%	22.7174%
07	0.0113%	0.0144%	0.0193%	25.6202%	20.9757%	22.3547%
08	12.8281%	10.4209%	8.1911%	40.4122%	33.0582%	45.2588%
09	4.4344%	5.8413%	2.8094%	30.6764%	30.2321%	30.5450%
10	16.8330%	28.4447%	23.8609%	33.4913%	28.7649%	35.0113%
11	16.6383%	16.4499%	17.5444%	31.9856%	25.3273%	35.1624%

Table 7 Left to right side liquid flow split in the distributor based on section 10 and 11 flow rates.

Section	100% full flow rate			35% turndown flow rate		
	baseline	Case 2x	Case 4x	baseline	Case 2x	Case 4x
Left side (11)	76.03%	74.31%	75.74%	50.41%	49.96%	49.48%
Right side (10)	23.97%	25.69%	24.26%	49.59%	50.04%	50.52%

The level of the flow field fluctuation for the 100 % and 35% flow cases is visualized by means of RMS fluctuating liquid volume fraction and RMS fluctuating liquid velocity magnitude contours on the cross-sections marked in Figure 17. Figure 29 shows the fluctuating liquid volume fraction and Figure 30 shows the fluctuating liquid velocity magnitudes. The fluctuations for the two flow rates are typical for the different choices of the interfacial relaxation time scale. The interface relaxation time scale parameter study shows that the flow rate fluctuations resolved on the mesh do not change drastically in a statistical sense. The strong imbalance in flow split for the 100% flow condition persists for the different values of interface relation time scale used in the sensitivity study. The strong levels of liquid flow fluctuations (with some downward flowing liquid) in the vertical legs for 35% flow conditions shows that that the interface relaxation time scale is nearly appropriate.

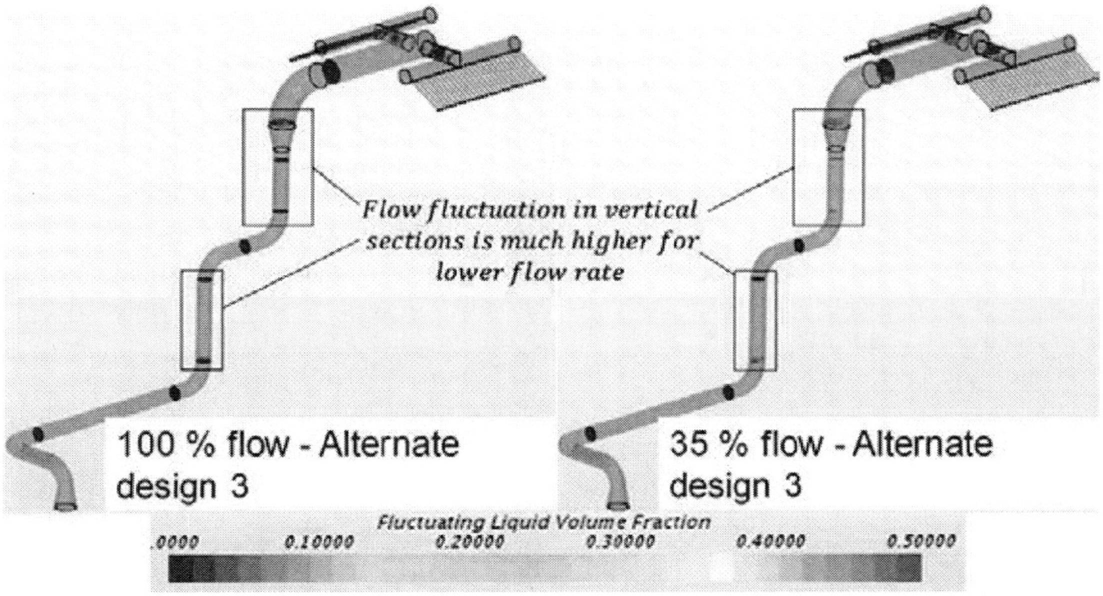

Figure 29 RMS fluctuating liquid volume fraction at various sections for the alternate design #3, 100% and 35% flow conditions (0.01 s interface relaxation time scale).

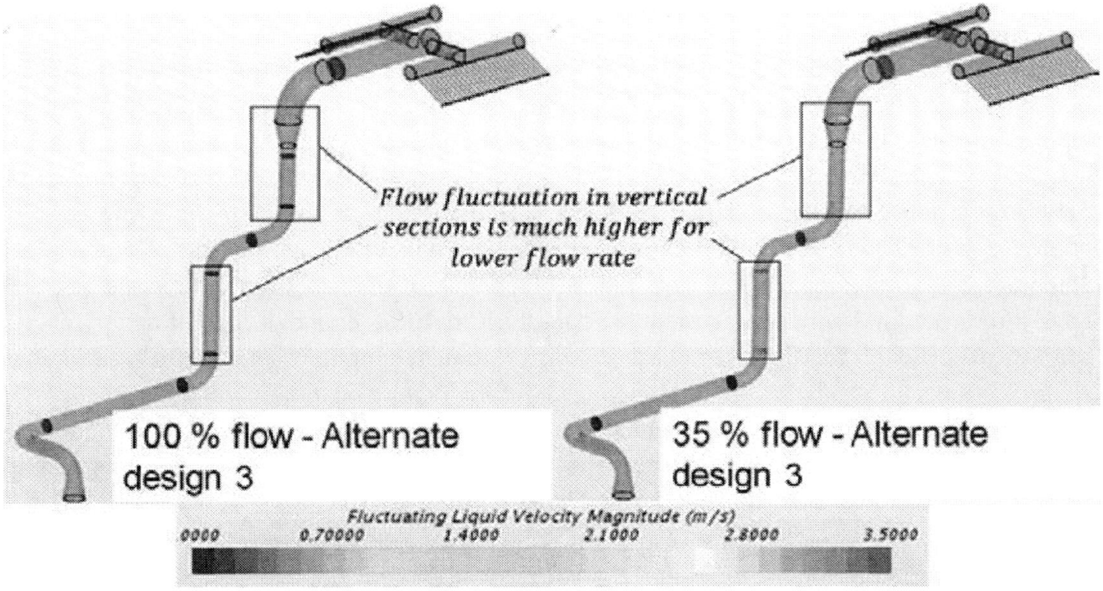

Figure 30 RMS fluctuating liquid volume fraction at various sections for the alternate design #3, 100% and 35% flow conditions (0.01 s interface relaxation time scale).

8. Grid, Time Step, and Iterative Convergence Study

To confirm that the CFD model results are not significantly affected by the choice of grid resolution, time step size, and chosen iterative convergence criteria at each time step, the 100% flow rate case for alternate design 3 was simulated on a finer mesh, with smaller time step size, and higher number of iterations per time step as shown in Table 8 using the same under-relaxation for all variables. The 100% and 35% flow case simulations used the same time step size and mesh, and thus the 100% flow cases would be most affected by discretization and iterative errors. The CFD model solution at a specific instant was interpolated to the fine grid and the fine grid was run further for comparison with the baseline grid solution. The comparison of the side to side flow balance coming out of the distributor (at the extruded tubes) is shown in Table 9 for a chosen instant. The grid study confirms the alternate design #3 side to side imbalance in liquid flow rates is not caused by numerical errors. The liquid distribution on the pipe walls for the refined grid is similar to the liquid distribution on wall for the baseline mesh.

Table 8 CFD model changes to assess the overall impact of numerical uncertainty.

	Total cell count	Number of prism layers	Wall adjacent cell height	Time step size	Iterations per time step
Baseline grid	804,475	2	$0.738\ mm$	$3.00 \times 10^{-4}\ s$	15
Refined grid	1,810,907	3	$0.273\ mm$	$1.11 \times 10^{-4}\ s$	20

Table 9 Assessment of CFD model numerical uncertainty for side to side flow balance coming out of the distributor (extruded tube outflows) at a particular time instant.

	Left side gas	Right side gas	Left side liquid	Right side liquid
Baseline grid	35.10%	64.90%	79.10%	20.90%
Refined grid	37.03%	62.97%	81.97%	18.03%

9. Conclusion

We make the following conclusions on the basis of this study:

- LSI modeling of two-phase gas-liquid flows is capable of handling internal flows in piping system with different flow regimes in different sections. CFD analyses of these flows is essential in many cases as the flow correlations and flow regime maps, are essentially applicable to idealized constant diameter horizontal and vertical pipes. The correlations are mostly based on certain fluid combinations and pipe geometries, and do not consider transient flow conditions.
- It is shown in this study that geometrical changes far upstream can have significant impact on the downstream flow. We see the large change in flow distribution in the 100% flow rate case for the alternate design #3 due to reconfiguration of bends upstream of the debutanizer.
- Some areas requiring further development are modeling of the relaxation time scale at the large-scale-interfaces, and robust treatment of two-phase outflow boundary conditions. Use of more sophisticated models for dispersed phase size distribution and momentum transfer is also beneficial.
- A verification and validation effort covering different fluids and flow conditions is necessary to establish high level of confidence in predictions. It should be mentioned here that if the computational model equations contain terms that are dependent on the grid size, the grid and time step refinement studies will need to consider the fact that the governing equations change with grid resolution.

10. References

1. A Large Scale Interface Multifluid Model for Simulating Multiphase Flows, H. H. Gada, J. Elias, M. P. Tandon, and S. Lo, Eleventh International Conference on CFD in the Minerals and Process Industries, CSIRO, Melbourne, Australia, December 7-9, 2015.
2. A Model for Predicting Flow Regime Transitions in Horizontal and Near Horizontal Gas-Liquid Flow, Y. Taitel and A. E. Dukler, AIChE Journal, v 22, Issue 1, January 1976.
3. Studies of Two-phase Flow Patterns by Simultaneous X-ray and Flash Photography, Memorandum No. AERE-M2159, G. G. Hewitt and D. N. Roberts, Atomic Energy Research Establishment, H. M. Stationery Office, Harwell, Berkshire, 1969.
4. STAR-CCM+ Version 10.02.010 CFD code User Guide, CD-Adapco, 2010.

480412 A Probabilistic Method for Quantifying Risk in Hydrate Forming Systems

Abstract:

As the development of oil and gas resources moves toward the deep water, the consequence of hydrate blockage increases. Risk avoidance by operating outside the stability region, or shifting the equilibrium curve using thermodynamic inhibitors, has been the dominant practise traditionally. More recently, economic pressures have led to a paradigm shift toward risk management strategies, where limited operation within the stability region is managed to prevent plug formation. To confidently employ these methods, predictive tools that incorporate the effect of hydrate formation on multiphase transportability and flowability of the system are required. Such tools allow for a deterministic calculation to assess the severity of hydrate formation for a given condition set. However, some properties, such as the hydrate nucleation temperature, are inherently stochastic, and others, such as the flowing bottomhole pressure, will inevitably fluctuate as the field is produced. Instead of choosing a representative value for these properties to perform a small number of deterministic calculations, this work presents an algorithm to explore the parameter space more broadly, by coupling a Monte-Carlo engine to a rapid steady-state flow framework. This permits flow assurance engineers to explore the sensitivity of their system to (i) hydrate formation, (ii) evolving field conditions and (iii) operating methods.

480414 Development of a Predictive Model for Hydrate-in-Oil Slurry Viscosity

Abstract:

The cooling of high-pressure oil and gas flowlines enables the formation of ice-like solids known as gas hydrates, which may be suspended in the oil phase, forming viscous, hydrate-laden slurries. As the slurry viscosity is a critical parameter to assess the probability of hydrate blockage, the development of a predictive viscosity model is critical for multiphase flow simulations. However, the current model used throughout the industry has received limited experimental validation. In this work, a controlled-stress high-pressure rheometer with a vane blade rotor was used to measure the viscosity of hydrate-in-oil slurries with two crude oils at watercuts of 5 to 30%. The dynamic viscosity profiles during hydrate formation and the steady-state flow curves were measured and compared to the current industry-standard viscosity model. The results show that hydrate slurries are shear thinning, and the magnitude of the flow curve increases directly with hydrate volume fraction. At steady-state, the deviation between experiment and the current viscosity model exceeds 50%, where the deviation increased directly with watercut. This result illustrates that a new generation of viscosity model is required to improve the accuracy of multiphase flow simulations. The yield stress of the fully-converted hydrate-in-oil slurries was measured after an eight-hour annealing process that was designed to simulate shut-in for a subsea flowline; the results showed that the yield stress increased exponentially with hydrate volume fraction.

Improving reliability of multiphase flow metering with thermodynamic models and correct field configuration

Xiaohong Zhang[1]*, Alessandro Speranza*[1]*, Merih Mustafa*[1]*, Richard Streeton*[2]
[1]*KBC Advanced Technologies plc, London, United Kingdom*
[2]*TechnipFMC, Rio de Janeiro, Brasil*

Abstract

With increasingly complex field developments and offshore exploration and production, multiphase flow meters (MPFM) have become very important in production optimization, reservoir management and fiscal allocation. The latest generation of MPFM can be also used in applications including well testing and surveillance, production allocation, artificial gas-lift optimization, production forecasting, process and equipment performance monitoring as well as flow assurance monitoring.

Since multiphase flow meters were introduced in the 1990s, the technologies have matured considerably. The industry recognises the importance of MFPM and their economic potential. With many years' experience, the industry has brought increased confidence in their accuracy and reliability. With correct field configuration, the advanced capabilities of MFPM can generate huge financial savings for operators in operational efficiencies, and most importantly in increasing overall reservoir recovery factors.

To achieve reliable and accurate measurements, MPFM need to be configured with accurate physical property (PVT) and thermodynamic models in order to accurately deduce the volumetric flow rates at operational conditions. The field configuration is also important in providing a consistent production process conversion to establish the correct volumetric flow rates at standard conditions.

This paper highlights the unique capabilities offered by the Multiflash software when integrated into the flow computer of the MPM multiphase and wetgas flow meter manufactured by TechnipFMC. The paper also demonstrates how very high accuracy gas phase properties can be provided by the thermodynamic models which are essential physical properties for flow metering at high gas volume fractions.

1 Introduction

Thanks to the increasing level of maturity and constantly improving performance, the technology of multiphase flow meters has been increasingly deployed throughout the oil and gas industry. Equipped with modern technologies, MPFMs nowadays offer many advantages and a high level of flexibility, over the conventional single-phase metering systems. Whereas the former can work across a range of flow regimes, the latter require the phases of the well streams to be fully separated at the point of measurement, with significant increase in cost of separating equipment. With the increase in offshore oil and gas exploration and production, MPFMs obviously play an even more important role in offering a cost-effective solution for well surveillance and monitoring, well testing, production optimisation, flow assurance and production/fiscal allocation metering (1).

The deployment of MPFMs affects several aspects of the field management. Well instability is a common problem during the life of field and during production decline. In many cases, it may be difficult to detect variations in flow rates from unstable wells using conventional separators; in such situations, MPFMs become a useful tool for providing detailed information and continuous monitoring of production rates that is critical for successful and effective field management. In general, production optimisation is essential in reducing production cost. This aspect is particularly sensitive for oil reservoirs requiring assistance with gas-lift. Once gas-lift is implemented, MPFMs can be used for finding the optimum gas-lift injection rate, as they are capable of instantaneously showing the oil flow rate as function of injection-gas flow rate. MPFMs also play important roles in flow assurance management, which is relevant to guarantee the flow of oil and gas from the reservoir to the sales or custody transfer point. Through more frequent or continuous measurement with MPFMs, it may be possible to identify potential blockages in the production system such as hydrates, asphaltenes, wax, sand and scale so that appropriate prevention and remediation strategies can be applied to control the problem. As we will show, when configured with compositional information, MPFMs can provide information on formation water breakout, allowing early assessment of options and effective decision-making.

Multiphase flow is a complex phenomenon and the performance and accuracy of the MPFM depends on the measurement technique applied in the meter, as well as the physical property (PVT)/ thermodynamic models used for field configuration. The aim of this paper is to highlight how PVT properties and thermodynamic models can enhance the performance of MPFM, and to discuss the most reliable approach to provide the accurate thermal and volumetric properties required by MPFMs using advanced equations of state models. The benefits of a dynamic and compositional approach in multiphase flow measurements will also be addressed.

2 MPM Meter Technology and Measurement Principle

The measurement principle behind the technology of MPFMs consists essentially in a direct collection of three quantities that characterise the multi-phase flow:

- Pressure gradient
- Flow speed of each phase
- Amount of each phase

With these three quantities, and having information on the density of the fluid and the flow regime, the flow meter provides enough information to reconstruct, in full, the flow conditions.

The MPM meter shown in Figure 1, employs a patented 3D Broadband™ dielectric measurement system (2) alongside a Venturi, a gamma-ray densitometer and advanced flow models (3 – 6) in the measurement of individual phase flow rates in both multiphase and wetgas environments. The 3D Broadband™ system uses high-speed electro-magnetic (EM) techniques for measuring the liquid/gas distribution, phase fractions, water-liquid-ratio (WLR) and gas-volume-fraction (GVF) within the pipe. The system is based on permittivity measurements performed through an EM wave frequency range of 20-3600 MHz over several planes within the meter sensor section. The system sweeps the EM wave frequency range 200 times per second, resulting in excellent sensitivity.

The Venturi creates a radial - symmetrical flow in the downstream 3D Broadband™ section, which would be the natural flow condition if the pipe were infinitely long. These flow conditions are ideal when using tomographic inversion techniques. The Venturi also produces the differential pressure required in determining the total mass flow rate. The gamma densitometer is simply a means of measuring the mixture density.

By combining the phase fraction measurements from the 3D Broadband™ with the total mass flow from the Venturi and gamma mixture density measurements, individual flow rates of oil, gas and water are determined.

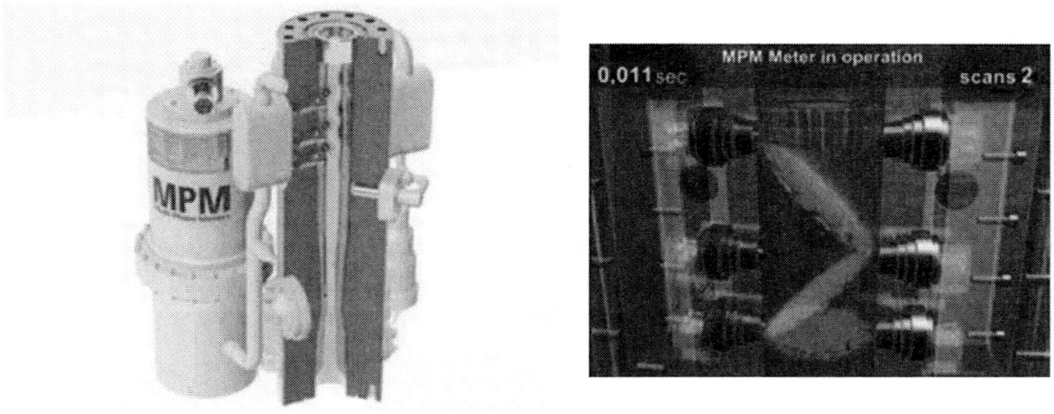

Figure 1 MPM multiphase meter constituent parts and 3D Broadband section

Through the Dual Mode functionality, the MPM meter is able to operate in both multiphase and wetgas environments and can switch between the two modes of operation automatically based on the instantaneous GVF. At ultra-high GVF's, where the liquid volume is very small compared to the total flow, the patented Droplet Count (DC) functionality significantly improves the measurement of the oil and water fractions. By using DC, the MPM meter can make precise measurements of minuscule liquid volumes in the GVF > 98.5% range where conventional technology is incapable of making true three-phase measurements. The DC method is also highly tolerant towards uncertainties in fluid properties. This is achieved through using the 3D Broadband™ rather than the gamma densitometer to measure the density of the overall fluid mixture. Further information on the MPM meter can be found in (8 – 20).

3 Field Configuration

'Field configuration' is a process by which the MPM meter is setup with data related to the phase properties of the oil, gas and water being metered. This information is required for two purposes:

- Prediction of the properties of individual oil and gas phase at the operating conditions of the meter. These include oil/gas densities, viscosities, the surface tension as well as compositions of individual phases, all of which are used at various stages in the flow calculation.
- To convert measured volumetric flow rates at operating conditions to reference conditions, typically standard conditions.

As part of a Joint Industrial Project and with participation from several major operators, the *Multiflash*™ PVT and physical properties software package produced by KBC Technologies has been embedded in the flow computer of the MPM meter. The integration of this advanced software into the MPM meter has several important advantages over the previous system of using lookup tables for field configuration. It also provides opportunities for new functionalities to be built using the compositional information that is now made available to the meter. The principal advantages of this method are:

- Direct upload of *Multiflash*™ (.mfl) PVT models to the meter for field configuration
- Reduced uncertainties from lookup table interpolation (21)
- Direct flash i.e. no limits in operating conditions
- Use of fully compositional advanced equations of state
- Account for presence of water in vapour form

- Real-time composition-based calculation of gamma mass absorption coefficients and permittivity
- Process-flash conversions to standard conditions

In traditional MPFMs technology, the density of the coexisting phases, at flow meter conditions is evaluated through an interpolation of the densities at surrounding conditions of pressure and temperature in preloaded lookup tables. With the embedded *Multiflash*™ software, a PVT model may be loaded directly to the meter so that lookup tables are no longer needed for fluid properties. At each calculation step, the embedded *Multiflash*™ software flashes the stored fluid composition at the actual conditions of the meter. The meter then has instantaneous information of the estimated composition of coexisting oil and gas phases at the meter's conditions.

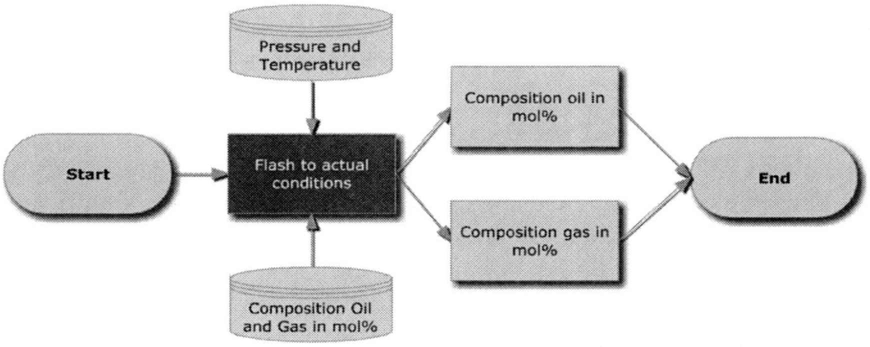

Figure 2 Flash of PVT model to actual conditions

The phase compositions are then used to generate the necessary fluid properties such as density, viscosity, surface tension, GOR etc.

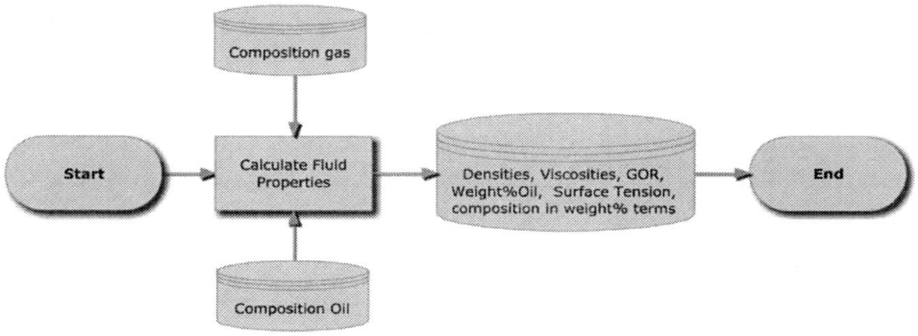

Figure 3 Oil and gas compositions used to generate necessary fluid properties

There are several advantages to this fully integrated and compositional approach. Firstly, users will have full control over the meter configuration. Indeed, engineers can create new PVT fluid models using *Multiflash*™ to incorporate the changes in the field to the new fluid models and send them directly to the meter. Secondly, there is no limit to the operating conditions associated with the configuration. *Multiflash*™ can provide the necessary fluid properties at all operating conditions, rather than being constrained to a limited pressure and temperature range, by the lookup table. Finally, interpolation uncertainties between entries in the current 5 x 5 tables is eliminated completely – this can have an important impact on the overall measurement uncertainty – particularly for subsea meters which typically operate over a large pressure and temperature range (21).

Through the utilisation of the embedded *Multiflash*™, the meter has an estimated oil and gas composition at every calculation step, which then provides additional useful information that can be used in the flow calculation and to provide further useful functionalities.

3.1 Advanced Equations of State

Using the *Multiflash*™ software, the MPM meter can make use of several equations of state (EOS) and thermodynamic models. *Multiflash*™ provides, in particular, several variants of RKS (Redlick-Kwong-Soave) (22; 23) and PR (Peng-Robins) (24), including volume shift (25) and various mixing rules, as well as advanced equations of state such as the CPA (26; 27) and the GERG-2008 (28). The GERG-2008 is an expanded version of the GERG-2004 (29) which is able to provide accurate predictions on the fluid phase behaviour, volumetric and thermal properties of natural gas mixtures containing 21 common components. The extended range of applicability of the GERG-2008 is up to 700 bar and 60 to 700 K. The GERG-2008 is particularly useful for high GVF gas applications where the accurate prediction of gas properties is important for accurately metering the small fractions of oil and water. Use of the GERG-2008 is therefore expected to reduce the overall measurements uncertainty of the meter at wetgas conditions.

3.2 Water Vapour

Previously, the MPM meter was only able to measure produced separated water. This could be made up of a combination of formation water and/or condensed water. Any water vapour would be measured as part of the gas as the meter has no way to discern between hydrocarbon gas and water vapour. Thus, when measured flow rates at actual conditions are transferred to standard conditions, the volume of water vapour that condenses as liquid would similarly be unknown.

There are therefore three areas that the embedded fully compositional *Multiflash*™ PVT software improves the meter performance in wetgas systems by providing reliable estimates of the following:

I. fraction of the liquid water at actual conditions that is condensed from vapour
II. fraction of water vapour in the gas at actual conditions
III. mass/volume of liquid water that condenses at standard conditions

The embedded *Multiflash*™ software can be also used to saturate the fluid in real-time to achieve the water vapour saturation at either:

- the actual operating conditions, or
- the reservoir conditions (real-time P&T from permanent gauges or fixed input)

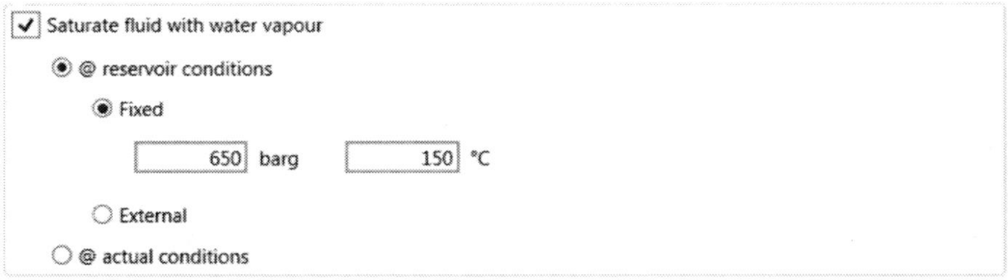

Figure 4 Configuration of water vapour saturation in the MPM meter

If water saturation is performed at actual conditions, then the meter will be able to predict II and III above, but not I. If the saturation conditions are reservoir P&T, then the meter will be able to predict all of I, II and III above.

In an ideal setup, a Modbus input can be provided to the meter of the reservoir pressure and temperature, such that the reservoir conditions are updated regularly. The embedded *Multiflash*™ PVT software will saturate the PVT model with water using the current reservoir conditions. The water-saturated PVT fluid model may then be used to:

- Provide gas and oil properties to the flow calculation, taking into account the presence of water vapour
- Provide the following additional output variables (at actual and standard conditions):
    - Water vapour mass fraction
    - Mass/volume flow rates of water vapour
    - Mass/volume flow rates of dry gas
    - Mass/volume flow rates of condensed water

The graphic below provides a description of how the water vapour calculation works internally using *Multiflash*™.

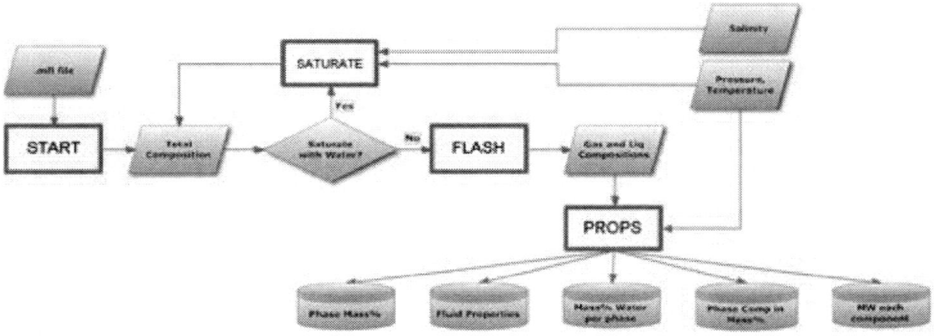

Figure 5 Water vapour saturation and property generation

To take full advantage of the *Multiflash*™ water vapour functionality, the fluid can be modelled using the "CPA" Equation of State which is a modified version of the "SRK" for use in water saturated systems and hydrate calculations. The CPA also includes a Peneloux volume shift to improve liquid density predictions. The CPA has an excellent performance in the prediction of water vapour saturation for natural gas mixtures.

3.3 Gamma and Permittivity Models

As described in (30), the measurement principle applied by the MPM meter is based on a composed approach to estimate the phase fractions in operations. Permittivity of the mixture is measured through the described 3D Broadband™ antennas at various frequencies. The relations

$$\varepsilon_{mix} = f(\alpha_o, \alpha_w, \alpha_g, \varepsilon_o, \varepsilon_w, \varepsilon_g) \tag{1}$$

are then inverted to have first estimate of α_o, α_w and α_g for the oil, water and gas phase fractions respectively. This estimate can then be refined, by means of a gamma densitometer measurement, which relates the density of the total fluid to the gamma photon count rate as follows

$$\gamma = N/N_o = e^{-d\mu\rho} \quad (2)$$

where N is the gamma count rate, N_0 is the empty pipe gamma count rate, μ is the mass absorption coefficient for the fluid, d is the diameter of the pipe and ρ is the density of the fluid, which is given by

$$\rho = \alpha_o\rho_o + \alpha_w\rho_w + \alpha_g\rho_g = \sum_p \alpha_p\rho_p \quad (3)$$

As per the above governing relationships, the field configuration must provide information on the mass absorption coefficients as well as the single-phase permittivity of each phase. For the hydrocarbon phases this can be handled using the embedded *Multiflash*™.

3.3.1 Mass Absorption Coefficient

The mass attenuation coefficient of a multi-component system is calculated using the weight% of each component along with the single component mass attenuation coefficients. i.e.

$$\sum_{n=1}^{80} \varphi_{C_nH_{2n+2}} * \mu_{C_nH_{2n+2}} + \varphi_{H_2S} * \mu_{H_2S} + \varphi_{CO_2} * \mu_{CO_2} + \varphi_{H_2O} * \mu_{H_2O} \quad (4)$$

where μ is the mass absorption coefficient of a single component and φ the weight% of that component in the fluid.

Using the known composition at the current operating conditions as predicted by the embedded fully compositional *Multiflash*™, the meter is now able to make real time corrections to the oil and gas mass absorption coefficients.

Importantly for gas fields, the presence of water vapour in the gas phase can also be corrected for by the embedded *Multiflash*™. The impact on the gamma densitometer performance can be important at high pressure where the gas may contain a significant volume of water in vapour form.

3.3.2 Permittivity

The permittivity of a multi-component system can be calculated using the Clausius-Mossotti relationship. The expression for the mixture permittivity for the oil or gas is:

$$\varepsilon_{mix} = \frac{2 * \beta_{mix} + 1}{1 - \beta_{mix}} \quad (5)$$

The polarizability, represented by β_{mix} will be proportional to the total number of molecules as represented by the total density and also proportional to the contribution from each component with a weight factor given by the individual weight fraction:

$$\beta_{mix} = \rho_{mix} \sum_i \chi_i \varphi_i \tag{6}$$

where i is an index representing the component number, χ_i is the polarizability factor and φ_i is the mass fraction of component i. The same relationship can be used for oil or gas.

Using the known composition at the current operating conditions as predicted by the embedded *Multiflash*™, the meter is now able to calculate the oil and gas permittivity in real-time using the phase composition. Importantly, the gas permittivity can be corrected for the presence of water vapour in gas fields which can have an important impact on the accuracy of the oil and water flow rate measurement.

3.4 Process Flash

In a standard setup, the MPFM measures the flow rates of oil, gas and water at actual conditions and transfers them to a reference condition (most likely standard conditions). This conversion is traditionally performed through a simple single stage flash and is based purely on the predictions of the PVT fluid model. Such conversions to standard conditions are, however, strongly *path-dependent.* The exact path taken by the oil and gas between actual and standard conditions has, therefore, an impact on the conversion result. The diagram below gives an example of a typical production process where the hydrocarbon phases are separated at the various stages.

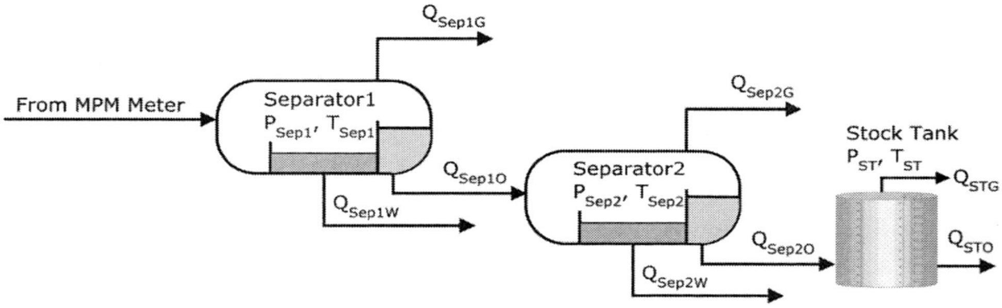

Figure 6 Example of a typical multi-stage flash process (3 stages)

There are several important observations in the process flash sketched above:

The difference between a single-stage and multi-stage process flash may be small for low pressures, low-shrinkage black oil applications. However, the higher the operating pressure and temperature of the meter and the more volatile the fluid, the greater will be the difference. For subsea meters, it is essential that a multi-stage process flash is used to convert measured flow rates as the operating pressure and temperature are generally high. Volatile fluids, such as gas condensates, change phase easily and, therefore, a process flash consistent with the field configuration should replace the traditional single stage flash.

1) The comparison between the MPFM measurement and the separator test can be performed by converting the MPFM flow rates to separator conditions or by converting both to standard conditions. Using the embedded *Multiflash*™, with the MPM meter, it is possible to output flow rates converted to both, separator conditions and standard conditions, reliably.
2) For subsea allocation and meter verification, it is often useful to convert the measurements from all subsea meters to the topside first-stage separator conditions as shown in the figure below. The totalised flow rates can then be easily compared to the first-stage separator at the

operating conditions. This method minimizes the PVT uncertainty due to process conversions and simplifies subsea meter verification.

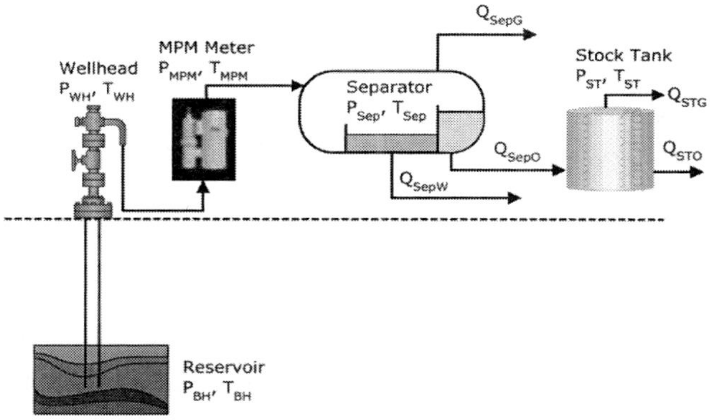

Figure 7 Conversion of subsea measurement to first-stage separator conditions

The meter can be configured with up to five downstream separator stages before standard conditions – which are sufficient for the majority of cases. Using the embedded *Multiflash*™ software, the MPM meter can also account for liquid condensation in the gas processing system. In gas fields there may be a significant liquid volume which condenses in the gas system. If this is the case, these liquids must be considered in the overall conversion to standard.

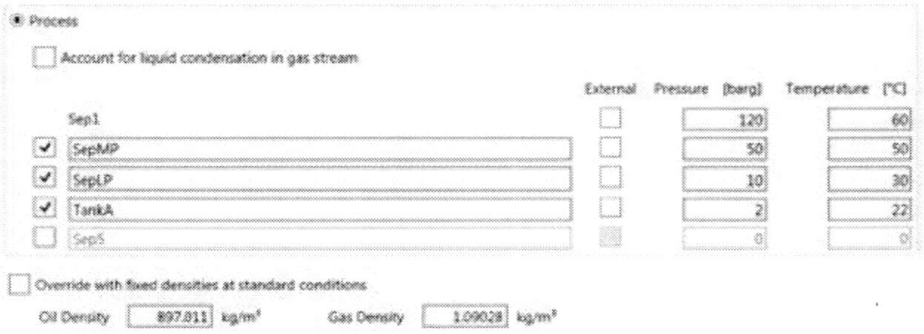

Figure 8 Configuration of the process-flash conversion in the MPM meter

4 Sources of Uncertainty

In using multiphase meters, factors such as field configuration and conversion to reference conditions must be handled with care. The graphic below shows how the overall system uncertainty is dependent on more than just the meter's measurement uncertainty.

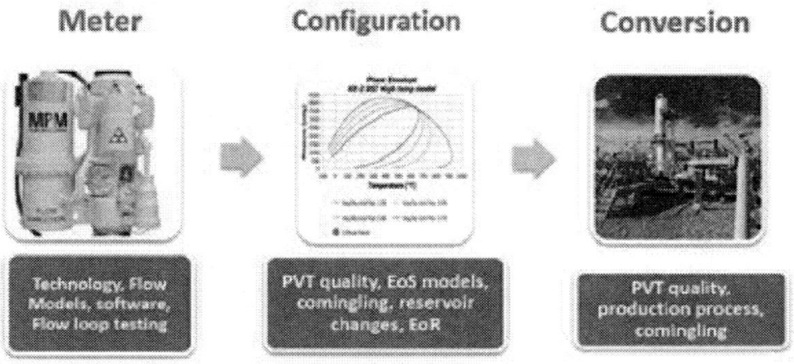

Figure 9 Elements of uncertainty in an allocation system

The accuracy of the field configuration is strongly associated with the following three main aspects.

- The uncertainty intrinsic in the PVT data which originates from the physical sampling process and analysis in the lab.
    - in some cases, only very basic or out-dated compositional information may be available
    - many different sampling methods exist whether bottom-hole, at surface or indeed more recently subsea, and each method has its own uncertainty
    - uncertainties are introduced though the lab equipment used, operator skill and may vary considerably between different labs
- The uncertainty that relates to the PVT package, characterisation and modelling process. There are many different commercial PVT packages available using a variety of EoS models and each has its own intrinsic uncertainty.
- The variations in the flowing fluid composition away from the base fluid composition during sampling, which means that the PVT fluid model no longer correctly predicts the phase behaviour of the well fluids. This may particularly occur at a later stage in the life of the well due to
    - the changes in the reservoir, or
    - the use of Enhanced Oil Recovery (EOR) techniques, or
    - wellbore effects, or
    - variations in contributions from individual zones in multi-zone completions

Measurements performed in real field applications have shown that the 'base' PVT uncertainty may be in the range of 1-3% or higher on oil and gas densities. If the well fluids contain more 'complicated' components such as H_2S and CO_2, the uncertainties may be much higher.

Where fields consist of a mix of black and volatile oils and/or gas condensate wells, which are comingled or gas-lifted, and where there is the use of gas injection in parts of the field, this may contribute to a wide range of produced compositions, leading to variations in the oil and gas density. These variations in composition may be unpredictable and result in large variations in the fluid properties, at operating conditions.

In the real world, uncertainties in the field configuration will always exist and must be managed correctly in order to obtain the best measurement results. In order to provide accurate measurements of all three phases, a multiphase meter should be adequately robust to handle such uncertainties. This is particularly important in applications where the operating point is close to the single-phase end-point of one of the phases, as shown graphically in the fluid composition map, in Figure 10.

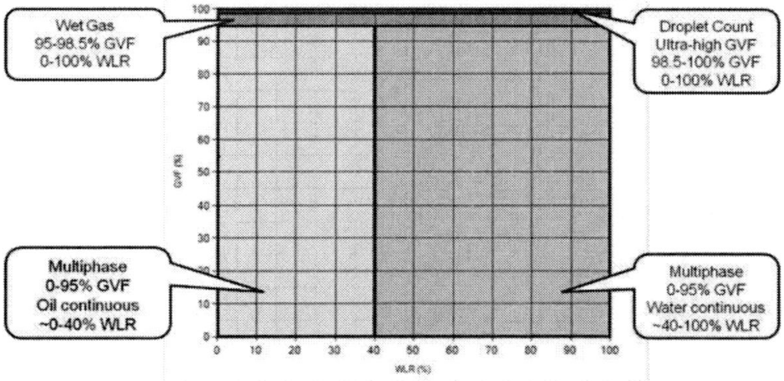

Figure 10 Fluid composition map

For applications in the red section of the map, gas properties are most important. For applications in the brown area, oil density will have a more important impact and similarly for the blue section where the user would be most concerned about the water properties.

Understanding the PVT sensitivity is of primary importance in evaluating the performance of a multiphase meter in real field conditions. The uncertainty specification of all multiphase meter technologies is subject to assumptions with regards to how well the PVT matches the true flowing fluids. As such, differences between configuration and real life fluid properties may cause a systematic bias in the measurement results, which cannot be accounted for, by simply performing longer accumulated well tests. In a fiscal allocation scenario, this is particularly troublesome, because systematic biases in measured results convert, directly, to misallocations between partners and governing bodies.

5 Uncertainty Analysis

Technological innovation and design improvement, as well as calibration through lab data contribute to reducing the effect of the uncertainty in the raw measurement of the flow meter, namely the flow speed and the flow rates of each phase. The effect of the PVT modelling is more complex and subtle however. Since the intrinsic uncertainty of the predictions depend strongly on the data utilised to establish the model, in order to minimise the overall measurement uncertainty, one needs to isolate the contribution of each one of the potential sources of uncertainty, stemming from the PVT configuration. By minimising each source of uncertainty and estimating its influence on the overall measurement, it is possible to compare the effectiveness and reliability of the chosen procedure and thermodynamic models to enhance the technology.

The following sections focus on the investigation of the ways in which thermodynamic models can provide an estimate on the uncertainty arising from imperfect knowledge of the feed composition, and from an incorrect conversion procedure to standard conditions.

5.1 Impact of density

As described in sections above, the measurement principles applied by MPFM technology and the meter performance are highly dependent on the PVT fluid models to provide the accurate phase fractions, compositions, and particularly the density of the individual phases as described in Eqs. 1 – 6.

In general the uncertainty of the phase fractions due to the density may be approximated (21) by

$$\left(\delta\alpha_{tot}^{\rho}\right)^2 = \left(\frac{1}{\rho_g^2} + \frac{1}{\rho_o^2} + \frac{1}{\rho_w^2}\right)\sum_j \alpha_j^2\left(\delta\rho_j\right)^2 \tag{7}$$

Thus, it is essential to minimise the uncertainty on the density of each phase, as the key physical property used by multiphase flow meters, in order to minimise the overall uncertainty on the flow measurements.

The embedded *Multiflash*™ PVT software offers a wide range of thermodynamic models including the well-developed CPA and GERG-2008 to accurately evaluate physical and transport properties of complex mixtures. In real gases modelling, phase density is regarded implicitly as a non-linear function of pressure, temperature and the feed composition. Thermodynamic modelling plays an essential role in providing physical properties of the phases in equilibrium. In the following, we will provide a detailed analysis on the uncertainty of phase density arising from various factors. Given its suitability (31; 32) to provide accurate partitioning of hydrocarbons in water and in presence of polar components (like alcohols or glycols), we will perform our analysis using the CPA-Infochem EOS (27) in *Multiflash*™, which is used as standard for calculations with water in the MPM flow meters.

5.1.1 The Feed Uncertainty

In normal operating conditions, the flowing fluid composition may vary with time due to various reasons. Since the fluid sampling is usually performed at the beginning of production, or earlier, during the appraisal and rarely repeated, such variations remain normally undetected. Such variations of the fluid composition away from the initial base fluid used for field configuration, may lead to the uncertainties of the predictions of phase densities.

In order to study the impact of feed uncertainty on phase densities, the base fluid composition is randomly perturbed with a Gaussian noise with zero mean and fixed standard deviation of 0.15 (15%). This allows the concentration of each component of the feed to variate randomly, with a normal probability distribution as shown in Figure 11, at most by 45-50 mole% in every direction (with 68% of the population within one standard deviation, or 15 mole% in each direction). By doing this, we generate a large pseudo-dataset of feed concentrations, effectively representing the possible casual variations of the feed and, therefore, its uncertainty. With such a data set, we can carry out a statistical analysis of the uncertainty on the phase density that originates from the uncertainty on the feed, and analyse it as temperature and pressure vary. By measuring the standard deviation of the distribution of the density of the phases corresponding to the perturbed feed compositions, we will have a measure of the uncertainty of the density, induced by the uncertainty on the feed. As we expect the density values to be more or less randomly distributed around the unperturbed value, we expect deviations distributed more or less normally, around 0.

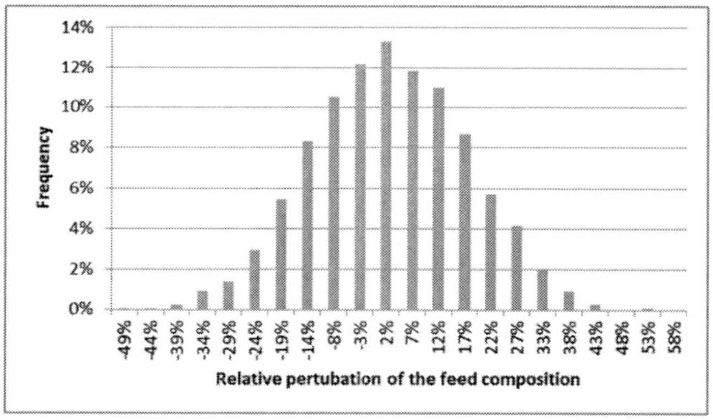

Figure 11 Distribution of the relative perturbation to each component in the feed

We consider two types of fluids in this study: a volatile oil with molecular weight of 61 g/mole and a wet gas with molecular weight of 19.2 g/mole, containing 90mol% of methane.

The phase behaviour of the two fluids predicted by *Multiflash*™ are presented in Figure 12. A concentration of 9 mole% of water is added to the two mixtures to have free water phase in the temperature range of interest. As shown in Figure 12, the volatile oil has a typical VLE phase envelope; the presence of water has small effect on the VLE line, where the water dew point line crosses the phase envelope. Similarly, the VLE of wet gas is affected by the water dew point. This is because the free water phase dissolves the gases and light hydrocarbons, leading to a shift of the VL phase boundary towards a lower temperature. Figure 12, also includes two isobars that show the conditions used in the analysis of the feed uncertainty.

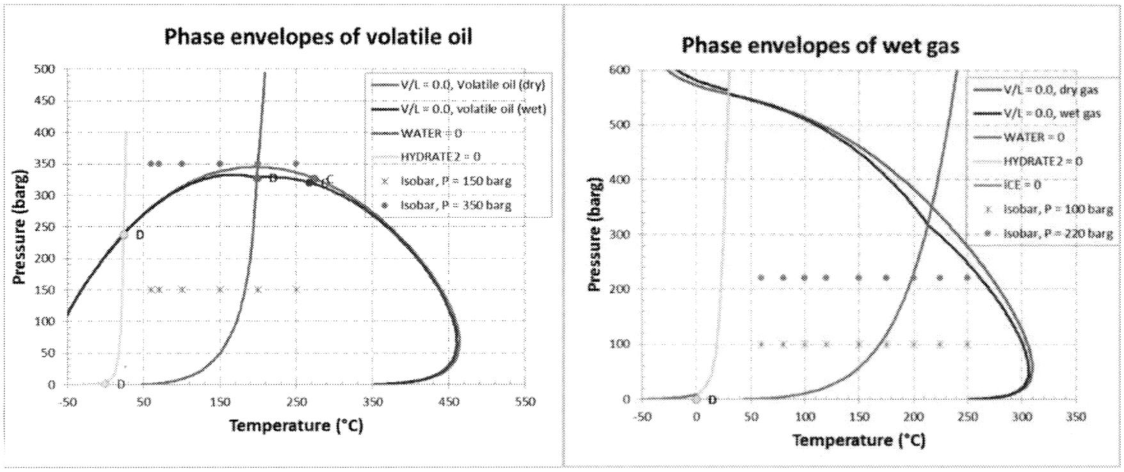

Figure 12 Phase envelopes of Vapour-Liquid, Water and Gas Hydrate II by Multiflash™

For the volatile oil, we consider two values of the pressure, at 150 barg and 350 barg, the former being below the saturation line and the latter being above it. The values of the temperature considered are between 60 degC and 250 degC as shown in Figure 12. With the composition perturbed as explained above and a given pressure of 150 barg, the density of the two phases vary within a distribution with standard deviation of approximately 3.5% for the gas phase and between 1.% and 1.5% for the oil phase. As expected, the density of the water phase is hardly affected by the feed perturbation, varying within a distribution with the standard deviation of 0.03%. The effect of temperature and pressure on the standard deviation of the phase densities at pressure of 150 barg is summarised in Figure 13 (left). Figure 13 (right) also shows the variation of the mean value of the density variations distribution (bias). The plot shows, as expected, no particular dependence on the temperature, as the density of the three phases vary around 0, more or less randomly. Note that the crossing of the water dew point line, between 100 and 150 degC appears to have hardly any effect on the density of the phases.

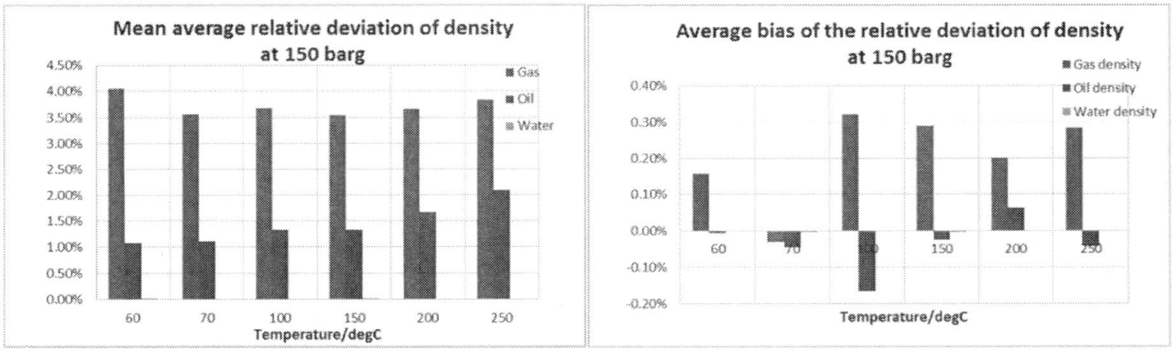

Figure 13 Standard deviation and bias of the relative deviations of gas, oil and water density as a function of temperature at 150 barg

For the case with pressure of 350 barg above the saturation line, the meter condition is located in the two-phase region with the oil and water phases. As mentioned earlier, the water density is not sensitive to the perturbation of the feed composition, but the oil density is affected with mean average deviation of 2.5%, which is slightly higher than that observed at the lower pressure. The bias (average value of the distribution) of the relative deviations of the oil density increases with temperature, whereas the standard deviation is not sensitive to the temperatures up to 150 degC when it gets slightly higher, as shown in Figure 14. This may be due to the fact that the conditions approach the phase envelope, the water dew point line and the critical region of the fluid, where the density of the oil phase is more sensitive to the variations of the feed composition.

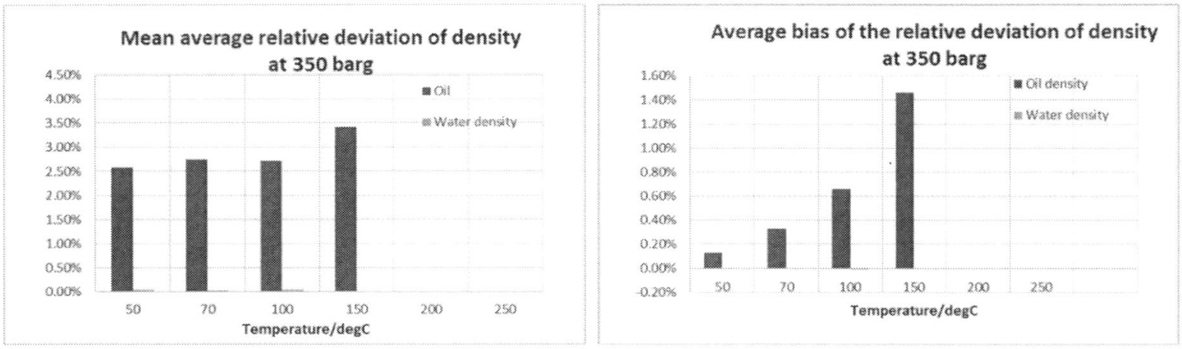

Figure 14 Standard deviation of relative variations of oil and water density as a function of temperature at 350 bar

The strong asymmetry in the fluid phase behaviour represented by the vicinity of the phase boundaries and the critical point affects the variation of oil density, introducing a bias that has the effect of a small "drift" in the average value. This, however, is related to the fact that within the population of the generated perturbed feeds, some compositions cause the conditions of temperature and pressure to switch between different areas of the phase diagram, between one- two- or even three-phase coexistence regions. The change in the density induced by such perturbations to the oil phase is, obviously very large. This effect can be appreciated by plotting the relative variations of the density of the oil phase, induced by the corresponding perturbations at 100 degC and 200 degC in Figure 15. The diagram shows that within the population of the perturbed feeds, a subset of compositions introduces a significantly large positive shift of the oil density, determining a drift of the average value. This effect is present only at 200 degC (green points), whereas at 100 degC (red points) the perturbed oil densities are distributed more or less symmetrically in the positive (above the horizontal axis) and negative (below the horizontal axis) semi-planes. The average density shift is, as a result, slightly positive, at 200 degC while it's roughly zero at lower temperatures.

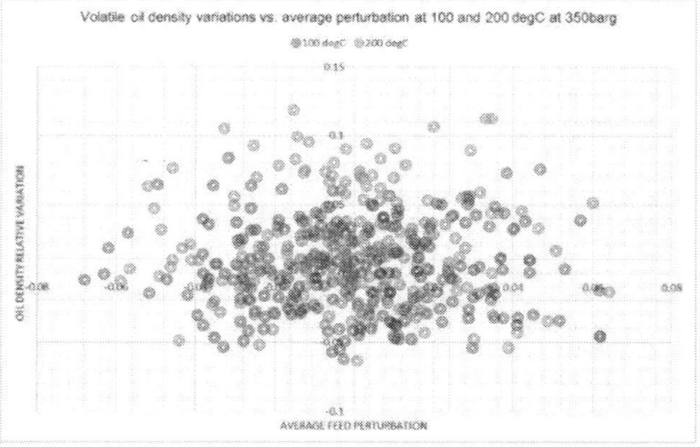

Figure 15 Oil density variation induced by average feed perturbation for two values of T, at fixed P

These data points correspond, in fact, to perturbations that take the pressure and temperature values in or out of a specific coexistence area and into a different one. As an example, in Figure 16 we report the phase envelope and water line of the unperturbed feed, for the volatile oil A, and the same lines generated by the perturbed feed corresponding to the maximum variation of the oil density recorded at 200 degC. In this case, for instance, the vicinity of the working conditions to the original water line is such, that the perturbation shifts the water line just enough to push the working point into the two phase region, whereas it was just inside the one phase region, in the original feed. As a result, in this particular case, the oil density varies by more than 15% (from 510 kg/m3 to 590 kg/m3) as the water dilutes light gases away from the liquid phase. By contrast, at 100 degC, the working point is far enough from the water line and the phase envelope, to induce, with the same perturbation, a change in the oil density about 11% (from 630 kg/m3 to 700 kg/m3), as the working conditions remain broadly in the same area of the phase diagram, albeit further away from the water dew point and the bubble point line.

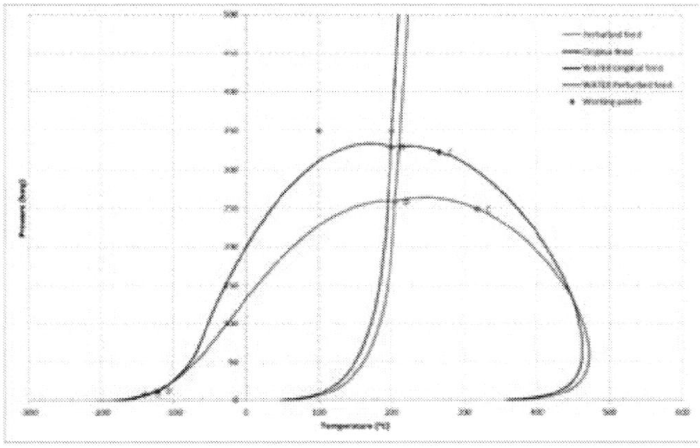

Figure 16 Phase envelope of the unperturbed oil and of the perturbed feed corresponding to the maximum variation in the oil density, induced by the perturbation at 200 degC

Such a drift in the oil density as shown in Figure 14, highlights another unexpected risk associated with the uncertainty on the feed composition. In some working conditions, the lack of appropriate knowledge of the feed composition may lead to conclusions about the density of the phases, which rely on the wrong assumptions of the phase coexistence. However, rather than introducing a source of uncertainty, this phenomenon may introduce a source of systematic inaccuracy on the value of the density, which, in this case, we estimated to be of the order of 1.5% (given by the average of the distribution of the density deviations). Rather than this being an error of accidental nature, i.e., a source *uncertainty*, it is a systematic source of *inaccuracy* of the measurement, which may affect all the flow rates and volumes measurements in one direction, leading, in this case, to a systematic underestimate (or overestimate) of the liquid density. This highlights, once more, the need of a correct field configuration, with appropriate PVT modelling, periodic checks of the working conditions against potential variations of the feed and, if appropriate, re-configuration of the flow meter, through upload of new field configuration data with updated PVT fluid model. The incorporation of *Multiflash*™ simplifies greatly this process, allowing rapid re-configuration of the instrument, through a direct upload of the fluid file, as well as greatly enhancing the capability to crosscheck the adherence of the fluid model and the predicted phase behaviour with the actual field evidence.

A similar analysis was carried out for the wet gas, to assess the uncertainty caused by the feed variations. The temperature range considered is the same as for the volatile oil, but the two pressures considered are 100 barg and 220 barg respectively. The results are summarised in Figure 17 and Figure 18.

As shown in the figures for the wet gas case, the standard deviation of the gas density distribution at pressure 100 barg is between 0.2% and 0.5% in the temperature range of 60 degC to 250 degC; much smaller than that of the volatile oil at pressure of 150 barg. However, the standard deviations of the oil density are much higher than that of the volatile oil with the deviation in the range of 2% to 6%. In Figure 17, it is also interesting to see a clear trend that the uncertainty of oil density increases with increasing temperature at 100 barg. This is to be expected, since as the temperature increases, towards the dew point line, the liquid phase has a composition which is increasingly dominated by some heavy species. Thus, when the composition change affects those species, this creates a big shift in the value of the density, contributing to widening the distribution. It is also interesting to see that while at 100 barg the standard deviation increases monotonically with the temperature, at 220 barg the width of the distribution increases at first and then it drops upon reaching the phase boundary. This is due to the fact at high temperature, the liquid phase in practice reduces to just the heavy end (C40) and methane, while at intermediate values of the temperature, there's a large presence of intermediate carbon numbers (C10-C30). Thus, at high temperature, the big shifts in the composition are only due to perturbed feeds with a significant variation of the amount of the two dominant components (C40 and methane), which, although present, are not very numerous. The distribution of the density values is, therefore, quite wide, but the standard deviation gets smaller. Note finally that the trend of the standard deviation is hardly affected by the crossing of the water dew point line at either values of the pressure. This is due to the fact that the water (H_2O) in the feed partitions between the gas and the water phase and is hardly present in the liquid phase. The formation of the water phase therefore changes the liquid phase very little.

For the density of the water phase, the finding is similar to what observed already for the volatile oil. As the water phase is dominated by water, with traces of other components, its density is almost insensitive to feeds variations (as long as the perturbed feed contains enough water to have a separate aqueous phase at the meter's conditions).

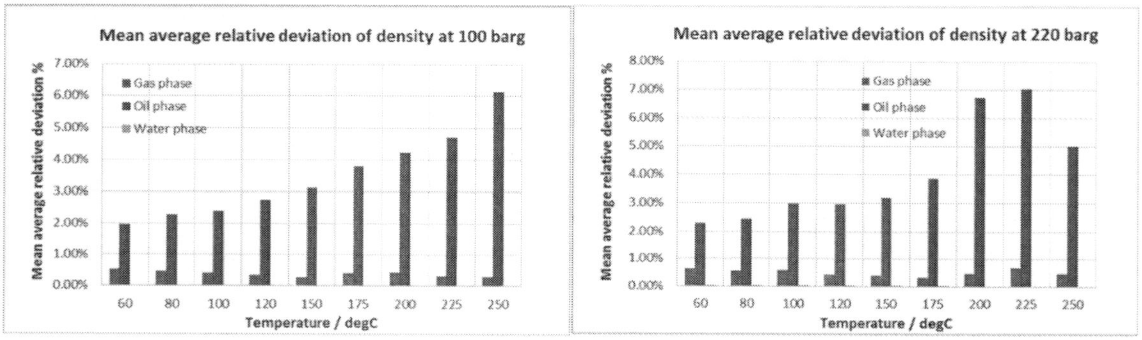

Figure 17 Wet gas: mean average relative deviation of phase density at 100 barg and 220 barg.

As mentioned above, 9 mole% of water was added to the wet gas to allow the formation of a free water phase at some of the meter's conditions. To assess the effect of the amount of water on the overall uncertainty on the phase density, this wet gas is also analysed with the presence of 1 mole% of water. The results are summarised in Figure 18.

As shown in Figure 18, the change of the amount of water in the system from 9 mole% to 1 mole% does not significantly change the trend of the deviations of gas and oil density in comparison with that of the case with 9 mole% of water. A further decrease of water down to 0.1mole% ensures that the water phase disappears completely in the temperature range of interest. However, the results found are also similar to those for the cases with 9 moel% and 1 mole% of water. This leads to strengthening the conclusion that this trend with temperature in the uncertainty of oil density is mainly governed by the presence of intermediate (C10-C30) and heavy components (C40) in the oil phase. When increasing temperature, the lighter hydrocarbons start to leave the oil to move into the gas phase and the resulting uncertainty drops. This, however, is visible only at high pressure, where the phase envelope is close to the highest value of the temperature (see Figure 12 right).

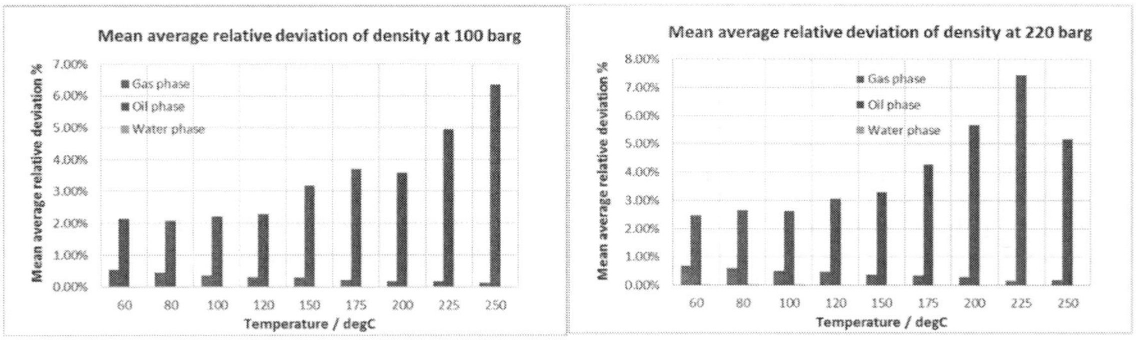

Figure 18 Wet gas with 1 mol% of water at 100 barg and 220 barg

This analysis shows that introducing an uncertainty of 15% on each component of the feed composition may lead to significant uncertainty on the density of the coexisting phases. This effect is quite significant in the gas phase of the volatile oil and the liquid phase of the gas condensate, where the uncertainty may reach 6-7%. In other cases, the uncertainty is, however consistently of the order of 1-5%. Furthermore, our analysis shows that in some conditions, the non-perfect knowledge of the composition may lead to systematic inaccuracies on the density of the phases, which are even more dangerous than the level of uncertainty since, if not predicted and accounted for, they would lead to a systematic error in the reading of the flow meter.

5.1.2 Choice of the EoS

The choice of thermodynamic models is also important in estimating the impact of uncertainty in density on the MPM performance. For cubic EOS, the use of the volume shift correction can greatly improve the liquid density predictions. The detailed analysis by Ruehle et al. (21) shows a very clear indication on how strong is the impact on the density of all the phases from the choice of EOS. The non-advanced versions, that do not contain the volume shift term, exhibit deviations from the average that reach the 20% of the actual value. On the other hand, an accurate choice of the variant of equation of state and of mixing rules can reduce the inaccuracy of the evaluation of the density to 2-3% for the density of the gas and give a consistent result, with deviations of the order of less than 1-2% for the density of the liquid and aqueous phases. Availability of density data to tune the EOS may help reducing the uncertainty even further. In the field configuration for the MPM meter, it is therefore imperative that the metering engineer accesses the correct set of models and EOS, to control and reduce the potential sources of error in the measurement.

5.1.3 Static lookup tables vs dynamic approach

The evaluation of physical properties based on temperature and pressure lookup tables is the most common in MPFM applications. However, this procedure can introduce further sources of uncertainty to the measurements, due to two main drawbacks:

- Uncertainty due to the variations of the fluid composition from the one used to generate the lookup table
- Uncertainty of the physical properties interpolated by the pre-calculated values on the temperature and pressure grid

As analysed previously (21), the latter could introduce a further 1-5% uncertainty on the density predictions of fluid phases, particularly for the condensates or volatile oils with large GOR or in fluids with high concentration of CO_2 and H_2S. As shown by Ruehle et al. (21) and pointed out in Sec. 5.1.1, if the flow meter conditions is near a phase boundary, the interpolation may introduce a further source of error to the measurement. The use of lookup tables greatly enhances the risks associated with the error introduced by the vicinity of a phase boundaries, as nearest-neighbour interpolation methods may lead to an interpolation between points that are in different regions of the phase diagram, causing the density of the phases to vary greatly and unpredictably from one point to the other and generating unphysical shifts in the values of the density or the physical properties. This, combined with the uncertainty on the actual location of the phase boundary, induced as seen in Sec. 5.1.1, by a non-perfect knowledge of the feed composition, could cause the evaluation of the physical properties to be persistently wrong. A rigorous flash calculation, based on a fully compositional approach, together with a continuous check of the feed composition, provides much more accurate and reliable method of calculation.

5.2 Conversion to standard conditions

Multiphase flow measurements are typically carried out at operating conditions. However, the measured values are conventionally reported to standard conditions, in order to provide values that can be compared directly with the separator measurements at the point of delivery. Therefore, the obtainment of the correct measurement at standard requires accurate information on how the fluid is processed downstream from the meter to standard conditions. This conversion typically involves a multi-stage process separation and must take into account the evolution of gas from hydrocarbon liquids (typically oil reservoirs) or the condensation of hydrocarbon liquids from gas (gas/gas condensate reservoirs), in order to provide a correct mass and volume conversion factor

Based on a complete mass-balanced and an ideal multi-stage separation, a numerical expression for the conversion for both hydrocarbon liquid and gas stream to the standard conditions may be expressed as

$$Q_{std}^{oil\_volume} = \frac{Q_{ac}^{total\_mass} \prod_{i=1}^{nstage} (1 - F_i^{gas\_mass\_frac})}{\rho_{std}^{oil\_mass}} \qquad (8)$$

$$Q_{std}^{gas\_volume} = Q_{ac}^{total_{mass}} (F_1^{gas_{mass}{}_{frac}} / \rho_{1,std}^{gas\_mass} + \sum_{i=2}^{nstage} F_i^{gas\_mass\_frac} / \rho_{i,std}^{gas\_mass} \prod_{j=1}^{i-1} (1 - F_j^{gas\_mass\_frac}))$$

where the oil shrinkage and liquid condensation depend on the temperature and pressure at separator conditions, and the number of separation stages involved in the actual production process. In some cases, however, such detailed information on the processing facilities downstream from the MPM meter to the final delivery point is not available. Instead, in order to estimate the mass conversion factor, conventionally a single-stage separation to standard conditions is applied. The actual field production involves, on the other hand, more complex processing facilities than the assumed single-stage separation. Such a simplification will surely add further uncertainty to the flow measurements at standard conditions. To reduce such uncertainty, a correct way to convert the measured flow rates to standard conditions is to perform a more complex multi-stage separation calculation to follow the actual production process as sketched in Figure 19.

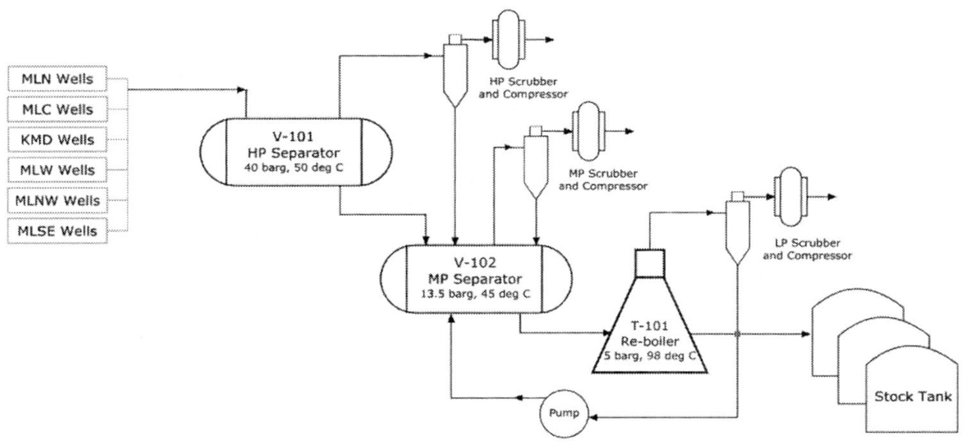

Figure 19 A simplified flow chart of a typical oil production process

In general, multistage separation processes are complex, but the temperature and pressure at the first primary separator conditions have the most influence on the mass conversion factor. To address the importance of the correct process separation, a 2-stage separation simulation as described in Figure 20 is applied to both the volatile oil and wet gas of the previous examples, whereas the conventional 1-stage separation simulation is treated as a base case for comparison.

The calculated results for the 1-stage separation are given in Table 1 and the results for the 2-stage separations for both volatile oil and wet gas are summarised in Figure 21 and Figure 22 respectively.

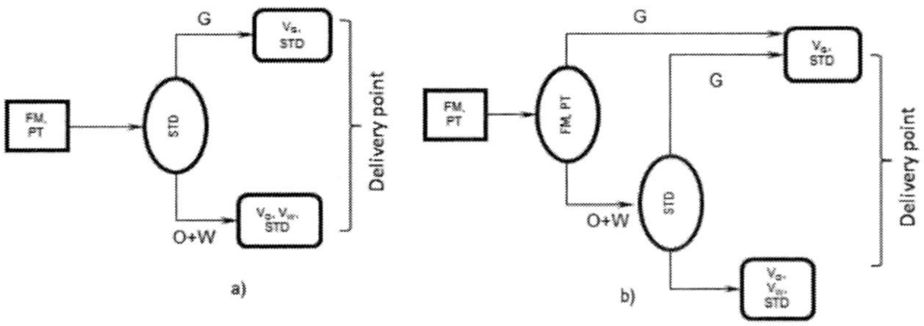

Figure 20 Schematic representation of the single- (a) and two-sage (b) separation transformation to simulate the facilities between the flow meter FM and the delivery point at standard conditions

	Volatile Oil	Wet gas
G/L mass ratio at STD	0.46	13.33
G/L+W mass ratio at STD	0.44	6.03
GOR (G/L) – scf/std	949	38242

Table 1 Single-stage flash at standard conditions for volatile oil and wet gas.

In Figure 21 and Figure 22, we report the mass ratio of gas to hydrocarbon liquid at standard conditions, as a function of the first separator pressure at a first separator temperature, together with the results from the single-stage separation. The figures clearly show that the difference of the mass ratio of G/L between the 2-stage and single-stage flash is large and is strongly dependent upon the pressure and temperature of the first stage of separation. Such deviations from the assumed single-stage flash, also vary with the type of crudes, as exemplified by the comparison between the volatile oil and wet gas.

Based on the 2-stage separation study, the uncertainty of the assumed single-stage separation for both the volatile oil and wet gas is mainly determined by the first separator conditions. For a given first separator pressure and temperature, the additional uncertainty on the G/L mass ratio introduced by the single-stage assumptions, can be estimated in 5.4% to 7.5% for the volatile oil, whereas the average deviation for the wet gas is much larger, in the range of 4.05% to 9.95%. The details of the mean relative deviations are summarised in Table 2. Once more, with the adoption of a rigorous and flexible thermodynamic calculation through the embedded *Multiflash*™, that allows the dynamic reconfiguration of the stages of conversion to standard conditions, the uncertainty can be controlled much more effectively than previously.

Temperature/degC	**80**	**50**	**35**
Rel. deviation of G/L mass ratio (Volatile oil)	5.40	6.71	7.59
Rel. deviation of G/L mass ratio (Wet gas)	4.05	6.07	9.95
Rel. deviation of G/L+W (Volatile oil)	5.43	6.74	7.64
Rel. deviation of G/L+W (Wet gas)	2.13	6.14	8.91

Table 2 Single-stage flash at standard conditions for volatile oil and wet gas.

As shown above, the thermodynamic models can play a very important role in improving the reliability of the MPFM applications. A sound PVT model allows, not only, the solution of complex phase equilibria, but also a reliable conversion of the measured flowrates of the phases to standard conditions. This study also highlights the impact of the first primary separator conditions on the overall oil recovery at standard conditions, indicating that process optimization is essential for maximizing production in the hydrocarbon production process.

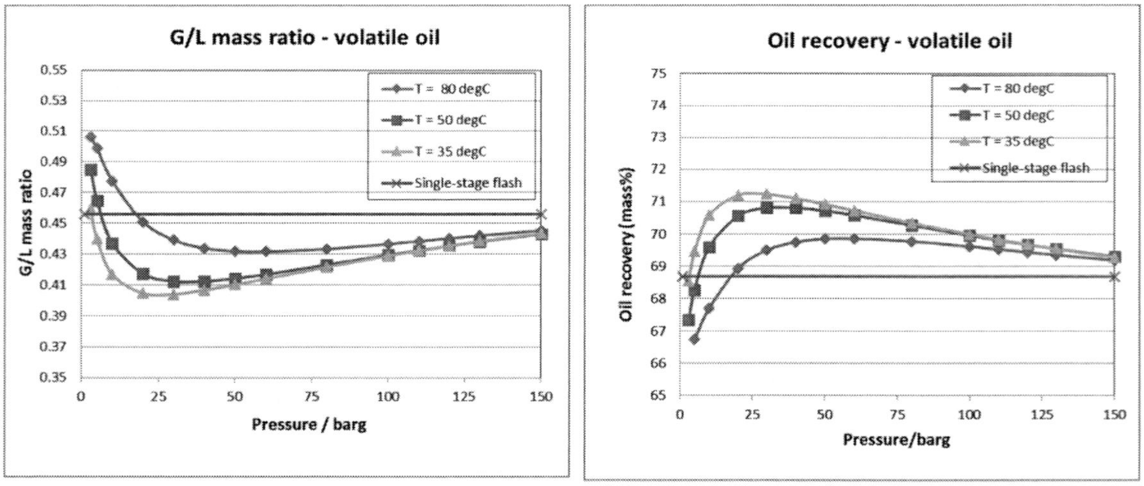

Figure 21 Volatile oil: G/L mass ratio and oil recovery mass percentage reference to standard condition

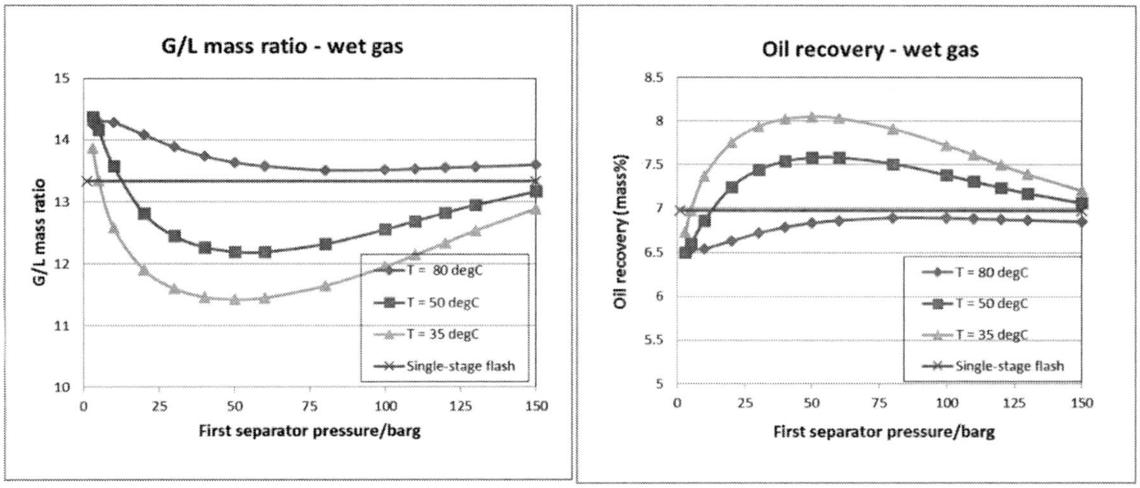

Figure 22 Wet gas: G/L mass ratio and oil recovery mass percentage reference to standard condition

6 Conclusion

The reliability of multiphase flow meter measurements mainly depends on the three aspects: the meter intrinsic technology, field configuration and conversion to standard conditions. Thermodynamic models play a crucial role on the last two aspects, in providing all the necessary physical properties for a correct flow rate calculation and feed into the first one. Among all the physical properties required by the MPFM, the phase density is the most influential and may have a significant impact on the system uncertainty of the meter. This study mainly focused on investigating the uncertainty of phase density arising from the variations of the flowing fluid composition and the incorrect field configuration and conversion. The detailed analysis was carried out with two very different type of fluids; a volatile oil and a wet gas.

To model the effect of variations of the flowing fluid composition, a Gaussian perturbation with a standard deviation of 15% was applied to all the components in the feed. The density variations induced by the perturbation on the feed composition were then analysed statistically. Our analysis shows that variations in the fluid compositions may lead to deviations of the gas density, which are roughly normally distributed, with zero mean. This translates, in practice, in an uncertainty on the gas density, of up to 4% in comparison with the original base fluid, which is induced by an uncertainty on the feed composition of about 15 mole% in each component. For the oil density, the uncertainty may reach 6-7% and shows a strong dependency on the temperature at fixed pressure, whereas the water density is not sensitive to the variations. The pressure too shows some effect on the relative deviation of both gas and oil phase. However, in cases in which the meter conditions are near a phase boundary or in the critical region, we noticed that small feed perturbations might translate in large deviations of the density, as the meter's conditions shift from one coexistence region to the other or as the critical point is approached. Indeed, in this case, we observed that the density deviations are not distributed normally around zero, but around a non-zero value, which increases as the meter conditions approach the phase boundary. By contrast with the previous observations, this in practice translates into the introduction of a source of systematic, rather than accidental, error, i.e., leading to a systematic inaccuracy of the density value that adds up to the uncertainty already encountered. In our case, this effect was estimated to, at most 1.5%. However, as density varies steeply across boundaries and across the critical regions, this effect might well be larger.

To overcome the difficulties and minimise the sources of uncertainty and potential inaccuracy of the measurements, we pointed out that an appropriate knowledge of the fluid sample and a sound PVT model is crucial. Furthermore, as unexpected variations of the compositions introduce both uncertainty and potential inaccuracy, a continuous check and if needed a re-configuration of the flow meter through an updated PVT model is strongly recommended.

The detailed knowledge of the production process downstream from the meter position to the delivery point is essential for accurate mass flowrate conversion. Assuming a single-stage separation, as common practice in flow meters applications, may lead to a large uncertainty in the estimated mass conversion factor, particularly for gases and gas condensates. Once again, the possibility to model more flexibly the facilities, through a more complex multi-stage separation, appears essential to overcome this potential source of inaccurate measurement.

This study highlights the importance of the thermodynamic models in improving the reliability of MPFM. The combination of advanced and innovative technological solutions, with the embedded *Multiflash*™ PVT software in the innovative and smart MPM multiphase flow meter represents a major step in the development of increasingly accurate solutions for fiscal allocation metering and advanced production management and monitoring.

7 References

1. *Handbook of Multiphase Flow Metering.* March 2005. Vol. Revision 2. ISBN 82-91341-89-3.

2. *Tomography powered multiphase and wet gas meter providing measurements used for fiscal metering.* Wee, A., et al. s.l. : NSFMW, 2007.

3. *Modelling Wet-Gas Annular/Dispersed Flow through a Venturi.* Maanen, M. van Werven and H. R. E. van. 6, s.l. : AIChE, June 2003, Vol. 49.

4. *Measurement of the Liquid Water Flow Rate Using Microwave Sensors in Wet-Gas Meters: Not As Simple As You Might Think.* van Maanen, H. R. E. s.l. : NSFMW, 2008.

5. *New Correction Method For Wet Gas Flow Metering Based on Two Phase Flow Modelling: Validation on Industrial Air/Oil/Water Tests at Low And High Pressure.* Couput, J. P., et al. s.l. : NSFMW, 2007.

6. *Liquid Correction of Venturi meter Reading in Wet Gas Flow.* de Leeuw, R. s.l. : NSFMW, 1997.

7. *Successful use and Implementation of Multiphase Meters.* Fosså, Ø. , Stobie, G. and Wee, A. s.l. : NSFMW, 2009.

8. *Measurement of Water in a Wet Gas.* Wee, A. and Scheers, L. s.l. : NSFMW, 2009.

9. *Multiphase metering with early detection of changes in water salinity* . Wee, A., Skjældal, I. M. and Bø, Ø. L. s.l. : Americas Workshop, 2009.

10. *Tomography powered 3-phase flow metering in the wet gas regime.* Bø, Ø. L., Wee, A. and Skjældal, I. M. s.l. : 8th International South East Hydrocarbon Flow Measurement Workshop, 2009.

11. *A combined multiphase and wet gas meter with in-situ measurement of fluid properties* . Wee, A. and Farestvedt, L. s.l. : Americas Workshop , 2010.

12. MPM. *Dual Mode.* White Paper 3.

13. —. *Salinity Measurement.* White Paper 2.

14. —. *In Situ Verification.* White Paper 8.

15. —. *Drople Count.* White Paper 7.

16. —. *Measurement Uncertainty Specification.* White Paper 1.

17. *A Methodology for Efficient Verification of Subsea Multiphase Meters used in Fiscal Allocation.* Streeton, R. and Blowling, I. Houston : Upstream Production Monitoring Conference, 2016.

18. *Advances in Multiphase/Wetgas Meter Technologies which aid the User in the Real World.* Streeton, R., Stobie, G. and Wee, A. Kuala Lumpur : South East Asia Flow Measurement Workshop, 2015.

19. *Campaign Well Testing of a Complex Reservoir using at MPM Multiphase Meter.* Streeton, R., Stobie, G. and Seidl, F. St. Andrews : North Sea Flow Measurement Workshop, 2014.

20. *Importance of PVT and Thermodynamic Models for Multiphase Flow Metering Applications.* Ruehle, Victor, et al. 2015. GPA Europe 2015 Annual Conference, Florence 16 - 18 September 2015.

21. *On the thermodynamics of solutions. V. An equation of state. Fugacities of gaseous solutions.* Redlich, Otto and Kwong, Joseph NS. 1, 1949, Chemical reviews, Vol. 44, pp. 233 -- 244.

22. *Equilibrium constants from a modified Redlich-Kwong equation of state.* Soave, Giorgio. 6, 1972, Chemical Engineering Science, Vol. 27, pp. 1197--1203.

23. *A new two-constant equation of state.* Peng, Ding-Yu and Robinson, Donald B. 1, 1976, Industrial \& Engineering Chemistry Fundamentals, Vol. 15, pp. 59 -- 64.

24. *A consistent correction for Redlich-Kwong-Soave volumes.* Peneloux, Andre, Rauzy, Evelyne and Freze, Richard. 1, 1982, Fluid Phase Equilibria, Vol. 8, pp. 7--23.

25. *An equation of state for associating fluids.* Kontogeorgis, Georgios M and Voutsas, Epaminondas . 11, 1996, Industrial & engineering chemistry research, Vol. 35, pp. 4310 -- 4318.

26. *Multicomponent phase equilibrium calculations for water–methanol–alkane mixtures.* Kontogeorgis, G. M., et al. 1999, Fluid Phase Equilibria, Vol. 158, pp. 201-209.

27. *The GERG-2008 wide-range equation of state for natural gases and other mixtures: an expansion of GERG-2004.* Kunz, Oliver and Wagner, Wolfgang. 11, 2012, Journal of chemical & engineering data, Vol. 57, pp. 3032--3091.

28. Kunz , O., et al. *The GERG-2004 wide-range equation of state for natural gases and other mixtures.* s.l. : VDI Verlag, 2007.

29. *Advances in Multiphase/Wetgas Meter Technologies which aid the User in the Real World.* Stobie, Gordon, Wee, Arnstein and Streeton, Richard. Kuala Lumpur : TUV, 2015. South East Asia Measurement Conference.

30. *Ten Years with the CPA (Cubic-Plus-Association) Equation of State. Part 1. Pure Compounds and Self-Associating Systems.* Kontogeorgis , G M, et al. 14, 2006, Ind. Eng. Chem. Res., Vol. 45, pp. 4855–4868.

31. *Ten years with the CPA (Cubic-Plus-Association) equation of state. Part 2. Cross-associating and multicomponent systems.* Kontogeorgis, Georgios M , et al. 14, 2006, Industrial & engineering chemistry research, Vol. 45, pp. 4869--4878.

COMPARISON BETWEEN DATA-DRIVEN AND PHYSICS-BASED MODELS OF A TABLETOP ANALOG TO AN OIL-WATER FIELD

Andrew Elsey, Ethan Myers, Ankush Agrawal, Kaiji Lu,
Jonathan Nguyen, Paul Saad, Matthew S. Elliott, and Gary L. Woods
Rice University, Houston, Texas, USA.

Xinli Jia, Olivier Germain, Keshava Rangarajan, Michael Redman,
Joseph B. Winston, Feifei Zhang, and Brent C. Houchens
Landmark-Halliburton, Houston, Texas, USA.

keywords: oilfield analog, data-driven model, neural network, physics-based model, compositional sensitivity

Abstract

A tabletop analog of an oilfield has been constructed to study strategies to mitigate water production. The tabletop analog represents a portion of an oilfield corresponding to a reservoir and producing wells. A porous layer contains a mixture of oil and water situated between non-porous layers, and miniature wellbore and wellhead assemblies extract the oil-water mixture. The reservoir pressure is created by either hydrostatic pressure or pumps, enabling the wells to produce. Sensors measure the pressures, total liquid flowrate, and oil/water ratio of produced fluids. The readings are validated by comparing them to the physical quantities of produced fluids.

The mitigation strategies require modeling that can predict production rates of oil and water. Both data-driven and physics-based models were developed to account for variations in reservoir pressure, oil/water ratio in the reservoir, production rates, and location of the completions relative to the oil-water interface. Two models of the reservoir were compared: a data-driven neural network and a physics-based model. The six variables included pressure at the oil reservoir, pressure at the water reservoir (psia), pressure at the wellhead, total volumetric flow rate at the wellhead, permeability multiplier, and produced water cut. Similar scenarios were run on the tabletop analog, and results were recorded to investigate whether or not the neural network was generalizable by comparisons of predicted and measured production rates of oil and water.

Background and Motivation

A 2009 study estimated that onshore US oil and gas fields alone produce 56 million barrels of water per day, including sources used to increase production, such as injected water [1]. Water remediation costs vary from as low as USD 0.07 to more than USD 4.00 per barrel, making water remediation a multibillion dollar per year industry [2]. Such high levels of water production greatly decrease the economic viability of oil fields. Consequently, tools used to diagnose high water cuts are important in maintaining well profitability. Currently, physics-based simulations are used to predict flow in the wellbore by numerically solving the conservation of mass and momentum equations. These are usually quite accurate, but require many simplifying assumptions, physical parameters, and measurements. Underspecified systems with estimated parameters are often used, but can produce inaccurate results. Machine learning algorithms are highly applicable in this area because the objective is to extrapolate behavior based on the known physical parameters and measurements of a reservoir. For example, operating pressures, flow rates, and water cut can be supplied to a machine learning algorithm to predict conditions in the reservoir that will lead to an unacceptable water cut. The accuracy of a physics simulation can be verified by evaluating its steady-state pressure, flow, and water cut values with the machine learning algorithm. This process can help to refine an underdetermined system to produce more accurate physics simulations and to detect anomalous behavior.

Models and Methods

This study develops three models of a mixed oil-water reservoir system: a physics-based simulator, a data-based simulator, and an experimental tabletop analog of a reservoir system. The following are held constant in the oil-water system: the reservoir permeability, production tubing size, temperature, oil and water viscosity and density (assumed incompressible). Six parameters that vary include the pressure of the oil reservoir (psia), pressure of the water reservoir (psia), pressure at the wellhead (psia), total volumetric flow rate at the wellhead (gal/min), permeability multiplier (dimensionless), and produced water cut (dimensionless). Collectively, these six parameters make up the input and output, which vary with the method of study. For the purpose of this study and to better mimic the tabletop analog, the water (bottom) and oil (top) reservoirs are physically separated by a horizontal barrier. Thus, there are two reservoir pressures. For steady state simulations and experiments, these are coupled. The hydrostatic pressure is the only variation between the reservoir pressures; consequently, they act as one connected reservoir. In the transient simulations, the pressure in the water reservoir is increased to simulate water injection.

Steady-state Physics-based Simulator

The conditions of the well were simplified to facilitate calculations for a one-dimensional flow of the oil-water mixture from the reservoir. Using conservation of mass and momentum equations, the production flow rates can be predicted and compared to the tabletop analog. Conservation of mass and momentum is implemented in the NETool™ simulator software. **Fig. 1** provides an example of the simulation setup.

A total of 18 realizations were performed by changing five of the six variables. The pressures at the oil reservoir and water reservoir were set, and varied with depth as given by the hydrostatic pressure increase shown in Fig. 1. At the wellhead, a target pressure and a target volumetric flow rate were also set. The permeability multiplier, which acts as a hyper-parameter to account for unknown details of the completions and near-wellbore region of the reservoir, was tuned to achieve these targets. A 1D, steady-state nodal analysis was then used to calculate the produced water cut, assuming that the oil and water flow was a homogeneous mixture with volume averaged properties of density and viscosity.

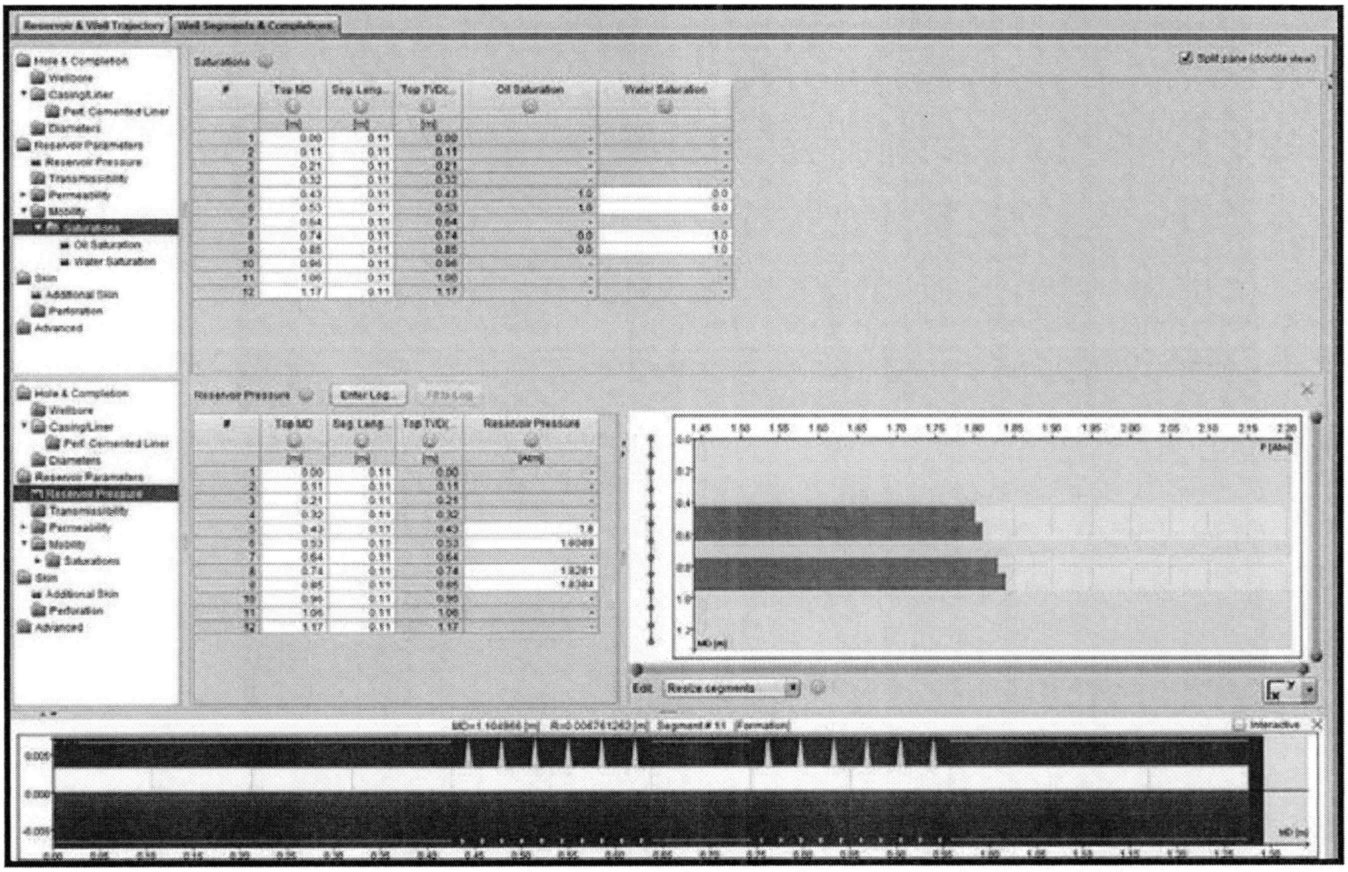

Figure 1. Setup of the physics-based simulator with water and oil producing zones.

Data-based Simulator

An artificial neural network was used to construct the data model. It uses the supervised learning paradigm, whereby a training dataset is provided for the model to infer the mapping implied by the input and output variables in the dataset. The training dataset is constructed to predict two scenarios. In both scenarios, the pressure of the oil reservoir, pressure of the water reservoir, pressure at the wellhead, and total volumetric flow rate at the wellhead are inputs. In the first scenario, only the produced water cut is predicted. In the second scenario, both the produced water cut and the input hyper-parameter (the permeability multiplier) for the underdetermined physics-based model are predicted. In both scenarios, a subset of the 18 physics-based simulations was used as training data.

The objective of the data-based model is to process limited input data (a subset of the six variables) through the neural network to predict the remaining unknown parameters. The data-based model can also improve the underdetermined physics-based model after the network has been trained based on fully-determined results of the physics-based models.

Two scenarios and associated neural network structures were designed. In scenario 1, the input layer 1 has five neurons (oil reservoir pressure, water reservoir pressure, wellhead pressure, total volumetric flow rate, and permeability factor). The hidden layers 2 and 3 have four and two neurons, respectively. The output layer 4 has one neuron, which corresponds to the predicted water cut. **Fig. 2** shows the network structure for scenario 1.

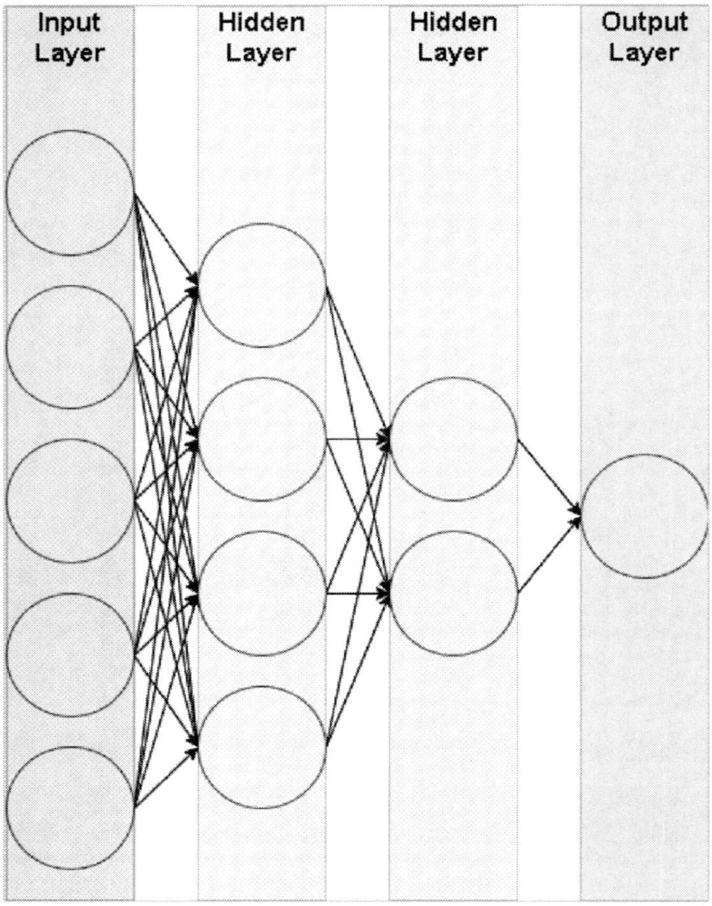

Figure 2. Fully interconnected neural network structure for scenario 1. Input data are entered to each input neuron, and the selected output data is produced by the output neuron.

In scenario 2, the input layer 1 has four neurons (oil reservoir pressure, water reservoir pressure, wellhead pressure, and total volumetric flow rate). The hidden layer 2 has three neurons, and the output layer 3 has two neurons, which correspond to the predicted water cut and the permeability multiplier.

For both scenarios, stochastic gradient descent optimizers were used. A mean squared error was used as the loss function. The model was trained with 50,000 epochs. To demonstrate the functionality of the neural net model in each scenario, it was compared against a simple ridge regression model. The ridge regression model is a standard linear regression model that minimizes prediction errors.

The only parameter that was tuned is the ridge regression parameter α. Because the purpose of the ridge regression is to establish a baseline of performance for the neural net, we set $\alpha = 1$. Given the small size of the training data set, the leave-one-out cross-validation paradigm was used for splitting the data, and the 18 trials of training and testing were averaged. This process enabled the best use of the data and provided the most unbiased result.

Tabletop Experimental Analog

The tabletop reservoir-analog serves as a scale model of a reservoir. A large-diameter, vertical PVC pipe is filled with a heterogeneous mixture of coarse to fine sand (as defined by the Udden-Wentworth scale) and represents the near-wellbore region of the reservoir. The sand serves as the porous media in which oil and water reserves reside. Two smaller diameter horizontal PVC pipes, also filled with sand, are connected to supply tanks, which together mimic the water and oil reservoirs.

Fig. 3 shows the system diagram of the facility. Oil and water are injected by means of hydrostatic pressure or pumps from source tanks. The reservoir contains porous material, and the pressure of each reservoir is measured. The oil-water mixture is extracted at the wellhead where pressure, volumetric flow rates, and oil-water composition are measured. All data is collected by the processing model. In future work, this model will be used to provide optimal control of oil extraction.

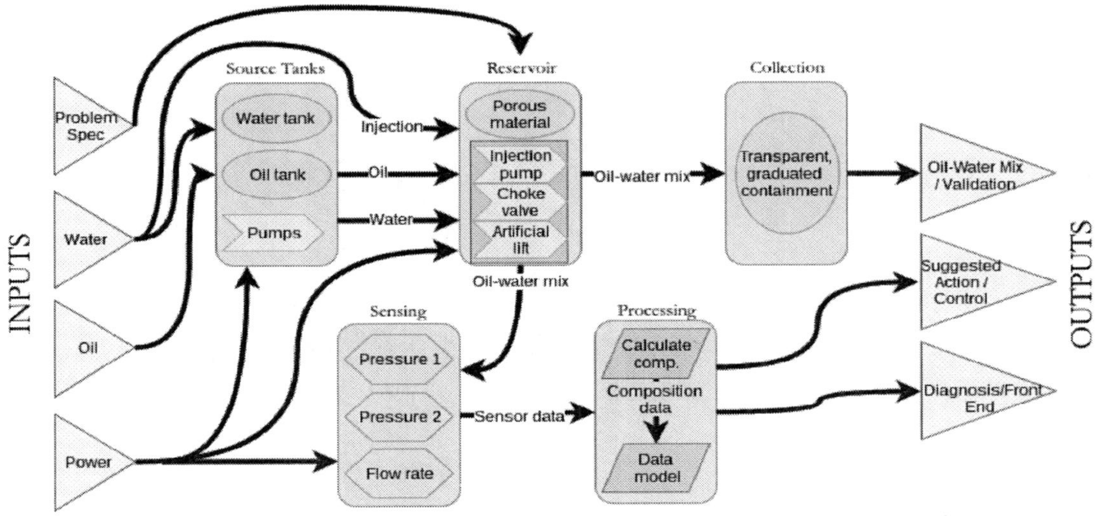

Figure 3. System diagram of the experimental tabletop analog.

The central column is 54 inches high and 6 inches in diameter. Caps on the top and bottom of the column mimic impermeable layers above and below the reservoir and enable pressurization. The central column includes three production well analogs with perforations at different depths into the column, which extract an oil-water mixture from the central column at their respective depths. Only one well is used during any experiment. The system can be expanded to incorporate multiple well systems in the future to observe the effect of multi-well interactions in a reservoir.

Fig. 4 shows a photo of the system. As shown in the figure, oil and water are introduced at the inlets under pressure, and the pressure sensors P0 and P1 measure the respective reservoir pressures. The oil-water mixture is produced through one of the production wells. Pressure sensors P2 and P3 determine the density of the oil-water mixture. The flow rate, pressure, and density of the mixture are logged through the entire testing process. **Fig. 5** shows a schematic cross section of the near-wellbore region. In this setup, each well has perforations at different heights. Production through the multiple completions can be studied by choosing a well with an appropriate perforation scheme.

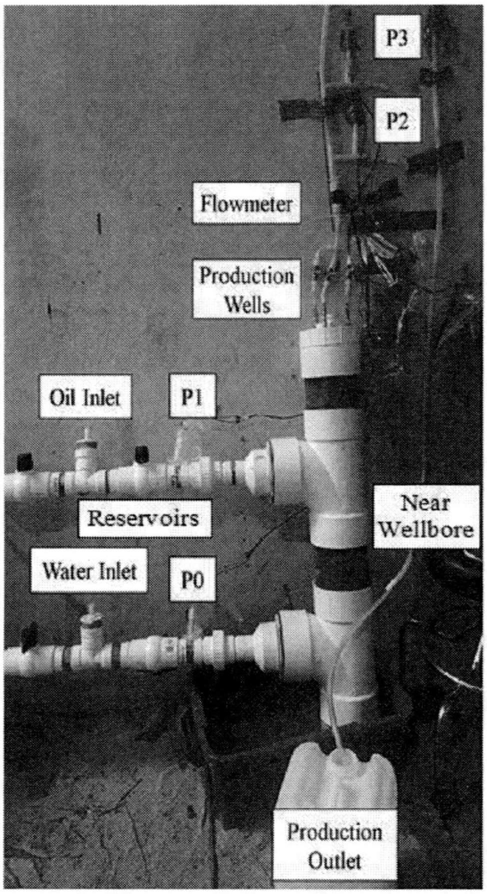

Figure 4. Photo of the tabletop analog.

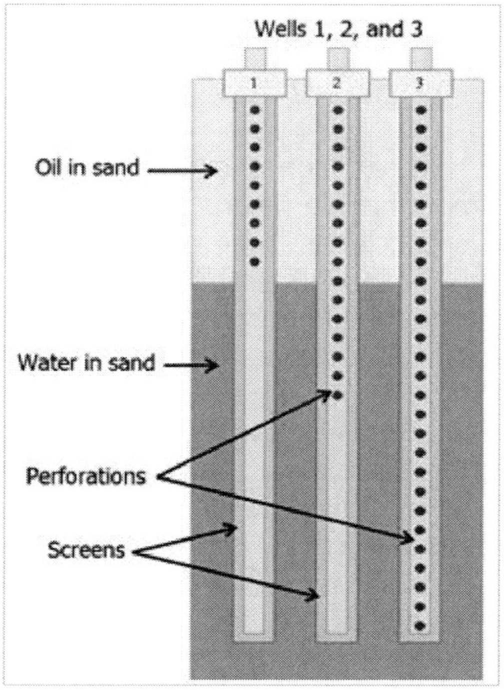

Figure 5. Diagram of multiple production wells and the near-wellbore region.

Oil is injected into the upper portion of the central column, and water is injected into the lower portion of the central column. The three extraction wells are controlled by ball valves, which can be manipulated to produce from the depth that will yield the smallest water cut at a given time within the reservoir analog. The oil and water pressures are controlled separately by hydrostatic pressure, varying up to 10 psi. The input pressure of the separate oil and water and the flow rate and density of the output oil-water mixture are monitored and logged. Two pressure sensors are used to determine the density of the output mixture. Because the density of each fluid is known, and the density of the mixture changes linearly with composition, the water cut of the production fluid can be easily calculated by measuring the density. This is useful to determine the effectiveness of mitigation processes.

Each of the sensors is connected to a Beaglebone Black development platform [3] running a Debian Linux operating system. The collected data is written by means of wi-fi to a database on a separate server, here an Intel® Edison development system [4] running a Yocto Linux operating system. The information is then processed through a series of queries to the server from any computer with an Internet connection. Each item in the database includes data type, identifier, value, and time stamp properties. This system infrastructure was designed to be modular.

Results and Discussion

Steady-state, Physics-based Simulations

Eighteen steady-state, physics-based simulations were run by changing the reservoir pressure, wellhead pressure, and a target of either the total produced volumetric flowrate (of oil and water together) or the produced volumetric flowrate of oil (independent of the water flowrate). **Table 1** lists the input values for these realizations. The temperature was set to a fixed 70°F throughout the wellbore to match the tabletop analog. The densities of water and dead-oil were constants of 999 kg/m^3 and 865 kg/m^3, respectively. The viscosities of water and dead-oil were nearly constant at 0.000975 Pa-s and 0.00357 Pa-s, respectively, over the pressures investigated. The produced oil and water was assumed to be a homogeneous mixture with volume-weighted average properties.

Table 1. Physics-based simulation inputs and outputs.

	Inputs				Outputs	
realization	Reservoir pressure at topmost connected node	Wellhead pressure	Total liquid (oil + water) volumetric flow rate at WH	Oil volumetric flow rate at WH	Volumetric fraction of water to total volume	Permeability multiplier
	[atm abs]	[atm abs]	[gal/min]	[gal/min]	[-]	[-]
1	1.3	1.0	0.05	--	0.783847	4.2880
2	1.3	1.1	0.05	--	0.782833	7.20318
3	1.3	1.2	0.05	--	0.777522	22.4863
4	1.5	1.0	0.05	--	0.784553	2.36988
5	1.5	1.1	0.05	--	0.784315	3.05269
6	1.5	1.2	0.05	--	0.783885	4.28817
7	1.5	1.0	0.08	--	0.784473	3.79703
8	1.5	1.1	0.08	--	0.784212	4.89297
9	1.5	1.2	0.08	--	0.783738	6.87821
10	1.5	1.0	--	0.5	0.701941	182.377
11	1.5	1.1	--	0.5	0.679084	259.534
12	1.5	1.2	--	0.5	0.640213	424.487
13	1.8	1.0	--	0.5	0.735318	92.4663
14	1.8	1.1	--	0.5	0.727513	111.304
15	1.8	1.2	--	0.5	0.716963	138.878
16	1.8	1.0	--	0.06	0.784171	7.96383
17	1.8	1.1	--	0.06	0.783976	9.20006
18	1.8	1.2	--	0.06	0.783711	10.8905

The outputs of the simulations were the produced water cut (volumetric fraction of produced water to total volume of produced liquid) and the permeability multiplier at steady state. The latter acts as a weighting factor to enable the target volumetric flowrate (total or oil) to be achieved at the given reservoir and wellhead pressures, if a solution is physically possible. It accounts for all of the unknown details about the completions, such as skin and losses. In this simulation, a single permeability multiplier applies to the completions in contact with both the water and oil regions.

Fig. 6 shows the variation of the permeability multiplier with changes in wellhead pressure, total target volumetric flowrate, and reservoir pressure. For fixed flowrate and reservoir pressure, the permeability multiplier must increase with increasing wellhead pressure to continue to achieve the target flowrate at a reduced pressure differential. Similarly, if the reservoir pressure is reduced with other variables held constant, the permeability multiplier must increase to achieve the same target flowrate. Finally, if the target flowrate is increased with other variables held constant, the permeability multiplier must increase. The output columns of Table 1 show the results for all 18 simulations.

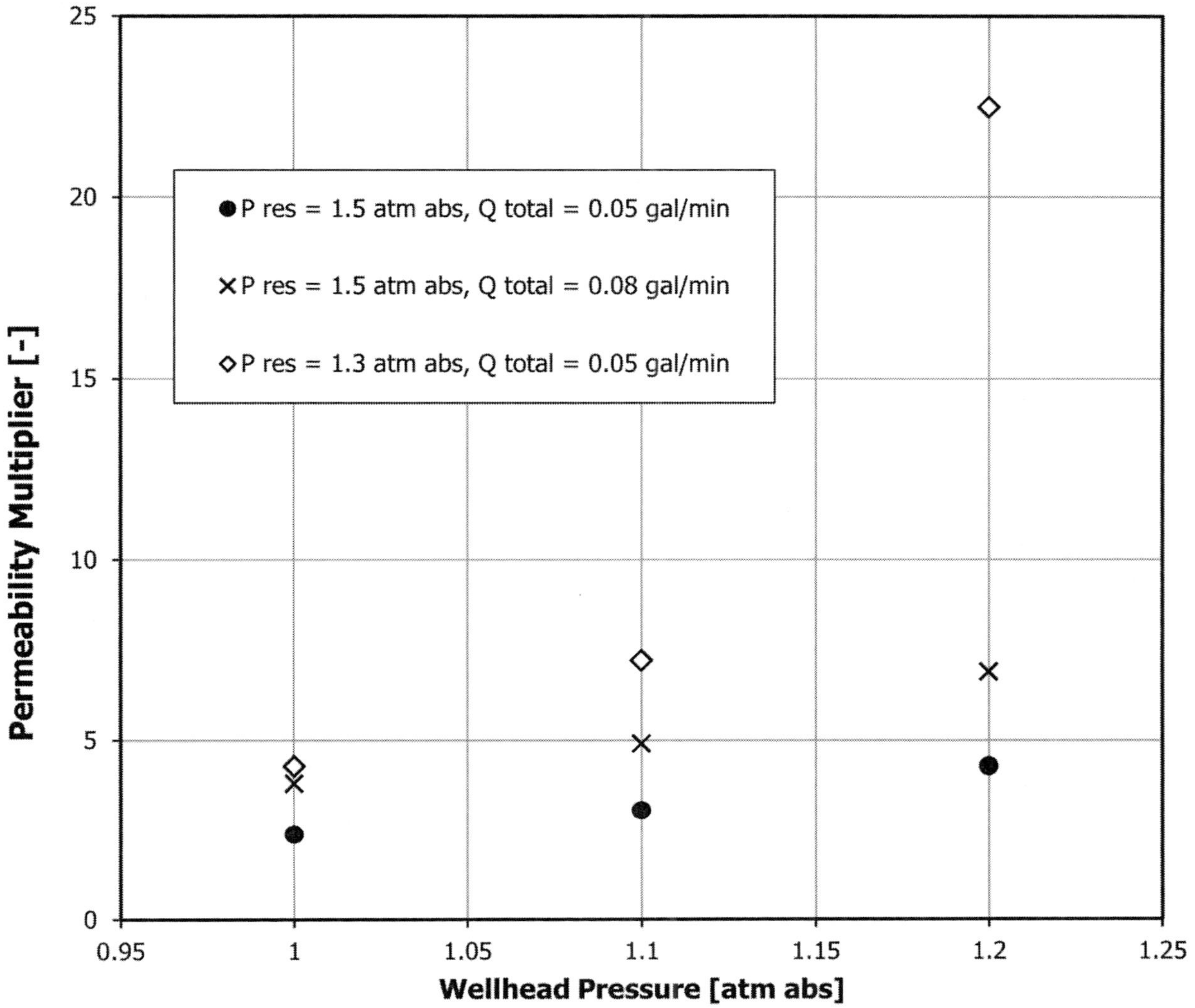

Figure 6. Permeability multiplier variation with wellhead pressure, total target volumetric flowrate Q_{total}, and reservoir pressure P_{res}.

Data-based Simulations Trained on Steady-state, Physics-based Simulations

The two neural network structures, representing the two scenarios, were trained and tested with a leave-one-out cross-validation paradigm on the 18 physics-based simulations listed in Table 1. Recall that scenario 1 had an input layer with five neurons, encompassing all of the physical-model inputs shown in Table 1; the reservoir pressure was separated into both an oil and water value, and the permeability multiplier was treated as an input. In scenario 2, the input layer had four neurons (oil reservoir pressure, water reservoir pressure, wellhead pressure, and total volumetric flow rate) and an output layer with two neurons corresponding to the predicted water cut and the permeability multiplier. **Table 2** shows the training and testing results of the neural network and linear ridge regression.

Table 2. Data-based training results for the neural network and ridge regression.

	Mean training time	Mean prediction time	Average mean percent error on predicted water cut	Average mean percent error on predicted permeability multiplier
	[sec]	[sec]	[-]	[-]
Scenario 1 with ridge regression	5.76e-4	7.46e-5	0.50%	NA
Scenario 1 with neural network	19.05	2.7e-3	0.36%	NA
Scenario 2 with ridge regression	6.3e-4	7.95e-5	2.2%	323.7%
Scenario 2 with neural network	17.19	2.1e-3	2.1%	25.83%

Scenario 1 demonstrates that, when the permeability multiplier is included in the inputs, the neural network configuration can predict water cut with less training time and with similar accuracy, as compared with a simple linear regression. However, the time for prediction is significantly longer.

Scenario 2, in which the permeability multiplier is a second predicted output, shows a similar trend for training time, prediction time, and accuracy of predicted water cut between the neural network and linear regression. Without the permeability multiplier as an input, the average mean errors on the predicted water cut increases somewhat as compared to scenario 1, but these are still well within acceptable limits. The neural network performs significantly better in the prediction of the permeability multiplier as compared to the prediction by linear regression.

Fig. 7 investigates the error on the permeability multiplier predicted by the neural network for each leave-one-out test. Points lying on the line indicate a perfect prediction. Most of the average error reported in Table 2 is associated with those realizations with the largest permeability multipliers. The worst errors in permeability multiplier prediction by the neural network are associated with either high flow rate cases, or low flow rates in which the difference between the reservoir and wellhead pressure is relatively small.

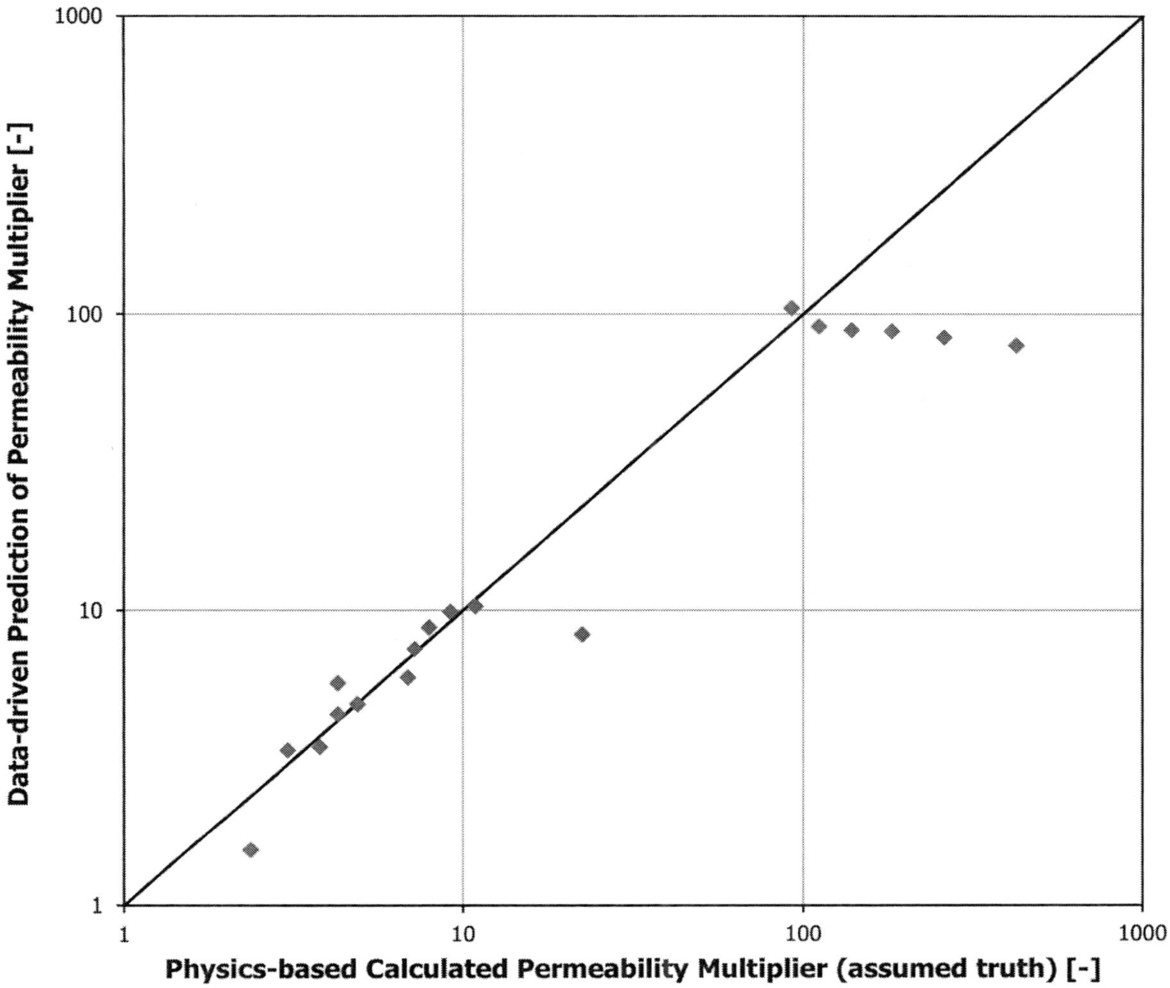

Figure 7. Predicted versus calculated permeability multiplier for the neural network in scenario 2, in which values falling on the straight line indicate an ideal prediction.

Tabletop Experiments as Compared to Steady-state, Physics-based, and Neural Network Simulations

Experiments were conducted with the tabletop analog by applying a hydrostatic pressure increase to the water reservoir. Three well completion levels (configurations) were investigated. The reservoir pressure was approximately 1.3 atm abs in experiments 1, 2, and 3, and approximately 1.4 atm abs in experiments 4, 5, and 6. Experiments 3 and 6 use well configuration 3, experiments 1 and 5 use well configuration 2, and experiments 2 and 4 use well configuration 1; Fig. 5 shows the three completion configurations. The results presented in **Fig 8** show that configurations with perforations deeper into the water saturated portion of the reservoir tend to produce much more water, as expected. On average, water cuts produced with the full length perforations (configuration 3 in Fig. 5) roughly reflect the predicted water cut from the NETool simulations, but the inability to reach steady state in the water cut makes such comparisons of limited value. Furthermore, in the NETool simulations, pressure and fluid supply comes from both reservoirs, not just the water reservoir, making comparisons even more difficult.

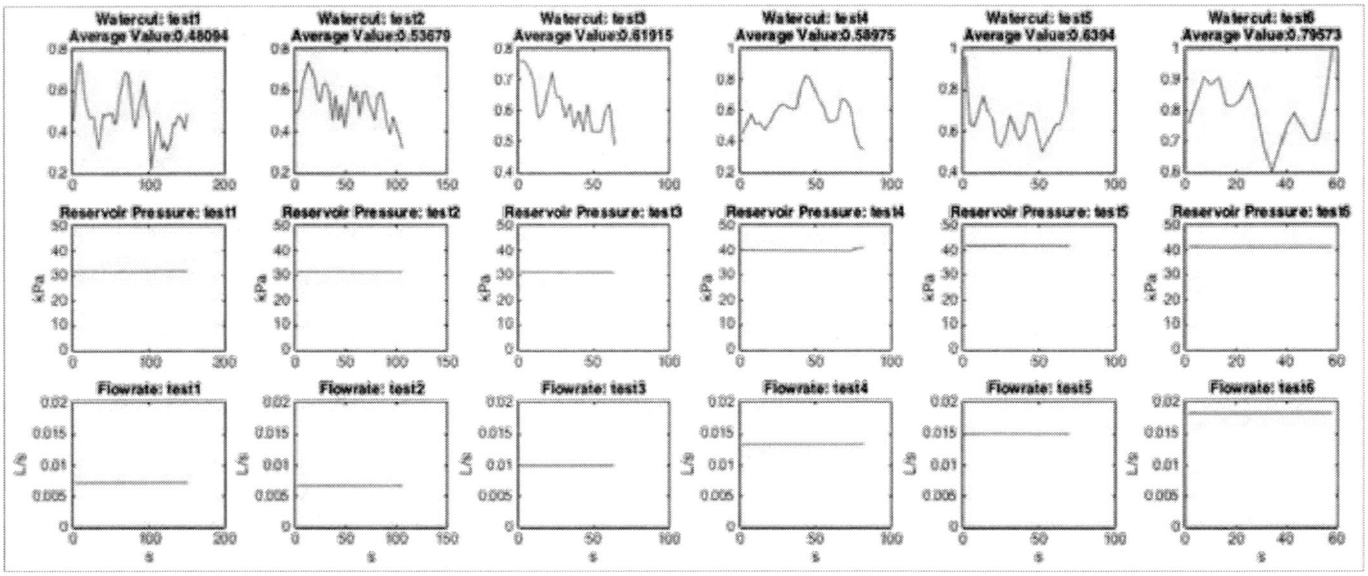

Figure 8. Produced water cuts and total liquid flowrates versus time for six tabletop experiments, for two nearly constant pressures in the water reservoir (~31 kPa g = 1.3 atm abs and ~41 kPa g = 1.4 atm abs) with completions shown in **Table 3**.

Table 3. Comparisons between water cuts from experiments on the tabletop analog and predictions from the neural network in the scenario 1 configuration.

		Tabletop analog experiment					
		1	2	3	4	5	6
Completion configuration		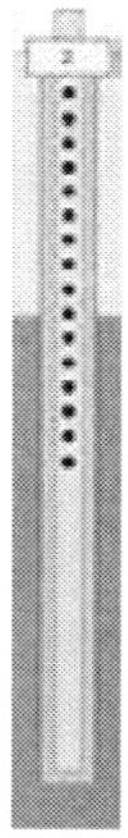	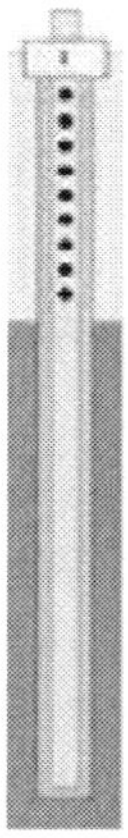	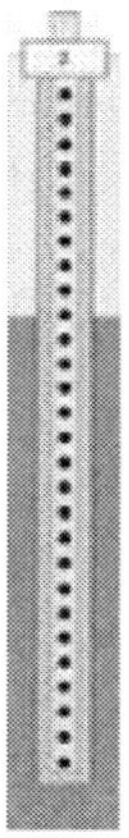	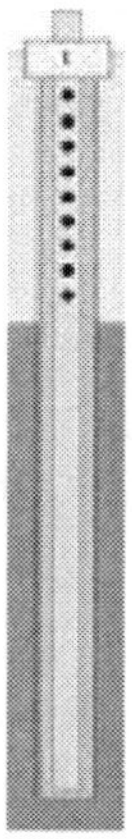		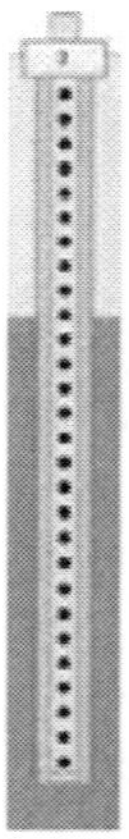
Water cut	Experimentally measured	0.4809	0.5368	0.6192	0.5898	0.6394	0.7957
	Predicted by neural network	0.4905	0.4875	0.4854	0.4860	0.4805	0.4791

The experimental data was also evaluated through the neural network in the configuration of scenario 1, trained by the NETool simulations, as shown in Table 3. Clearly, the predicted water cut values differ significantly from the empirical water cut. This is largely expected to be a result of the inability to achieve steady state in the experiments, which warrants further investigation.

Future Work

Several improvements are planned for the tabletop model. The current system uses hydrostatic pressure, which limits the range of pressures that can be applied to the system. A pump is planned to be added to future iterations to ensure better control of the system. Other upgrades will include flow meters with higher resolution and custom printed circuit boards to support all sensors and enable easy scalability.

Conclusions

A neural network was developed and trained. It demonstrates that produced water cut can be predicted by data-driven methods. The neural network is capable of modeling produced water cut better than a simple linear regression when both were trained on the same physics-based simulations. The long-term goals of this project are to predict missing parameters of underdetermined physics-based models by fully training the neural network on fully determined models and to eventually use data from the physical model that parallels the results of physics models to train the neural net. Each of these provide a different means of characterizing reservoir dynamics in the effort to reduce produced water cut.

References

1. Clark, C.E. and J.A. Veil (2009), Produced Water Volumes and Management Practices in the United States, Argonne National Laboratory report ANL/EVS/R-09/1 for the U.S. Department of Energy.
2. GAO (2012), Energy-Water Nexus: Information on the Quantity, Quality, and Management of Water Produced during Oil and Gas Production, U.S. Government Accountability Office.
3. Beagleboard.org®. BeagleBone Black. https://beagleboard.org/black (accessed 7 March 2017).
4. Intel® Edison Compute Model. https://software.intel.com/en-us/iot/hardware/edison (accessed 7 March 2017).

481508 Experimental Study and Molecular Thermodynamic Modeling of Winsor Type III Systems in Enhanced Oil Recovery (EOR)

Abstract:

Conventional modeling approaches for surfactant-oil-water systems in the "surfactant flooding" enhanced oil recovery (EOR) technique involve empirical correlations that require considerable experimental data and are not predictable for new systems [1]. Conversely, a straightforward molecular thermodynamic approach has been successfully used to describe the phase behavior of a variety of self-assembled systems including pure and mixed nonionic/ionic micelles and vesicles [2-4]. The ideal EOR formulations are surfactant-oil-salt water-alcohol mixtures that forms microemulsions (Winsor type III systems), which have an ultralow interfacial tension allowing them to reach deep into reservoir capillaries [1]. We show in this work how the key variables including alkane carbon number, salinity, surfactant type including tail alkyl chain length and head group characteristics, surfactant and alcohol concentration, and temperature can be used within the Blankschtein framework to predict attractive new W III systems. This study investigates three model n-alkane oil components (C_6, C_8 and C_{10}), sodium chloride as the salt, polyoxyethylene glycol monoether nonionic surfactants, and n-alkanol co-surfactants. Phase behavior formulation scans, in which one of the key variables is systematically changed in a series of prepared mixtures allowed to settle in graduated cylinders, are performed to identify formation and extent of the microemulsion phase for several combinations of variables.

[1] Salager, J.-L., Forgiarini, A. M., and Bullon, J., How to Attain Ultralow Interfacial Tension and Three-Phase Behavior with Surfactant Formulation for Enhanced Oil Recovery: A Review. Part 1. Optimum Formulation for Simple Surfactant-Oil-Water Ternary Systems, *J. Surfact. Deterg.*, **16**, 449-472 (2013).

[2] Puvvada, S., and Blankschtein, D. Molecular Thermodynamic Approach to Predict Phase Behavior and Phase Separation of Micellar Solutions. I. Application to Nonionic Surfactants, *J. Chem. Phys.*, **92,** 3710-3724 (1990).

[3] Yuet, P.K., and Blankschtein, D. Effect of Surfactant Tail-Length Asymmetry on the Formation of Mixed Surfactant Vesicles, *Langmuir*, **12**, 3819-3827 (1996).

[4] Shiloach, A., and Blankschtein, D. Measurement and Prediction of Ionic/Nonionic Mixed Micelle Formation and Growth, *Langmuir*, **14**, 7166-7182 (1998).

Distillation Tower Inspection for Turnarounds and Revamps

185 Kister Distillation Symposium 2017

Doug Bouck
Technical Consultant
Separation Solutions, Ltd

Safety

- Personal
  - Basic Refinery/Chemical Plant Safety Training
  - Confined Space Safety Training
  - Hole Watch Communication
    - Voice or Radio (Compatibility Issue)
  - Know Your Own Limits
    - Heights
    - Closed In Spaces
    - Darkness
    - Flexibility
    - Conditioning

2

Safety

- Hazards
  - Climbing
    - Outside: Cluttered Platforms and Ladders
    - Inside: Cords, Hoses, Damage
  - Encumbrances & Hang-ups: Harness, Meters, Radio, Tools, Coveralls
  - Falling Objects: Tray Parts & Tools
  - Trays & Other Internals: Loose, Corroded, Displaced, Sharp Edges
  - Partner or Other Workers/Inspectors
  - Tight Spaces – Exit Plan

3

Safety

- Hazards
  - Dust, Scale
  - Old Equipment - Asbestos Gaskets
  - Pyrophoric Dust & Scale Producing SO_2/H_2SO_3 Over 104F
  - Packing Fire Prevention:
    - "Causes and Prevention of Packing Fires", Chemical Engineering, p. 34 (July 2007).
- YOU are Responsible for YOUR Safety
- YOU are Responsible for OTHER'S Safety

4

5

Typical Tower Internal Manways

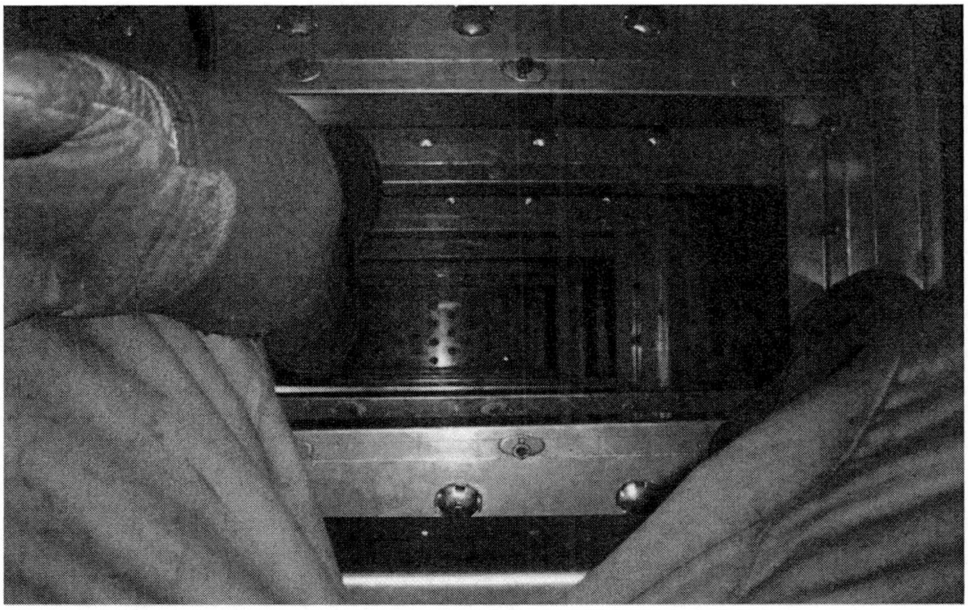

6

Tight Spaces

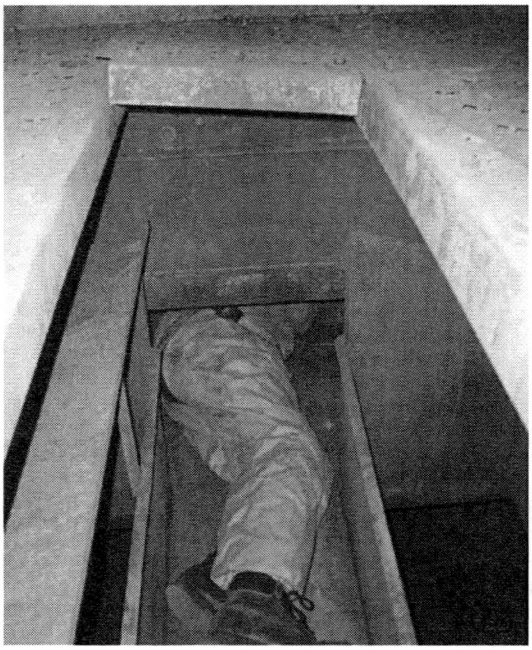

7

Multiple Personnel

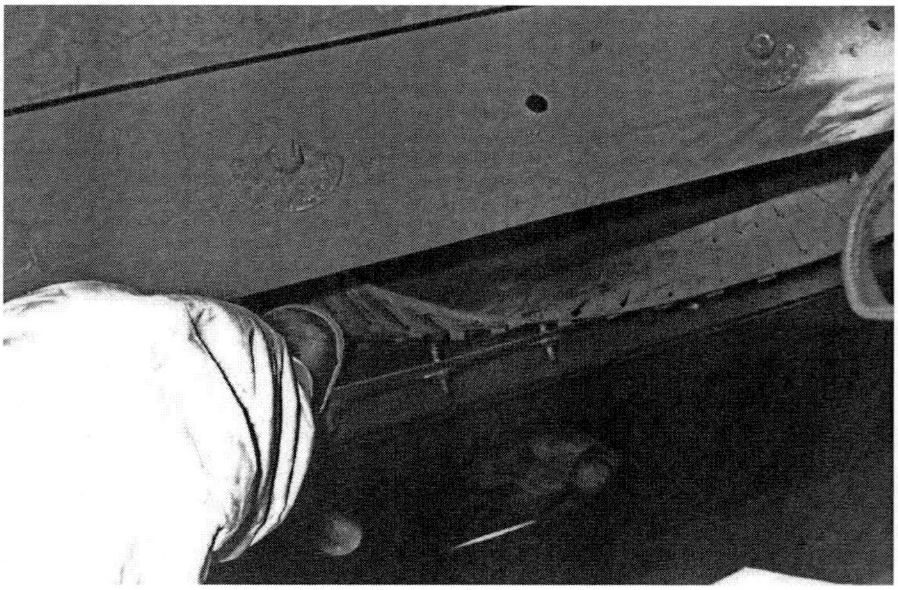

8

Tight with Tray Damage

9

Cluttered Manway & Internals

10

Scaffold Inside Tower

11

Workplan Responsibilities

- Process Inspection
- Maintenance/Mechanical Inspection
- Vendor Reps./Inspectors
- Contractor Inspectors
- Operations Inspection
- Training Inspection Tours
- Punch Lists and Change Orders
- Follow-up Inspections
- Signoff Responsibilities

12

Inspection Equipment

- Typical in Waist Pack
  - Lock-out Lock and Tag
  - Zip Lock Bags & Magnet
  - Soapstone + Yellow Crayon + Black Marker
  - 10-25 ft Tape Measure
  - 9/16 Wrench & Plyers
- Other
  - Shin Guards and/or Knee Pads
  - Note Pad, Drawings & Inspection Plan
  - 2 Light Sources with Fresh Batteries
  - 2+ Pens/Pencils
  - Camera (Hot Work Permit)

13

Waist Pack Contents

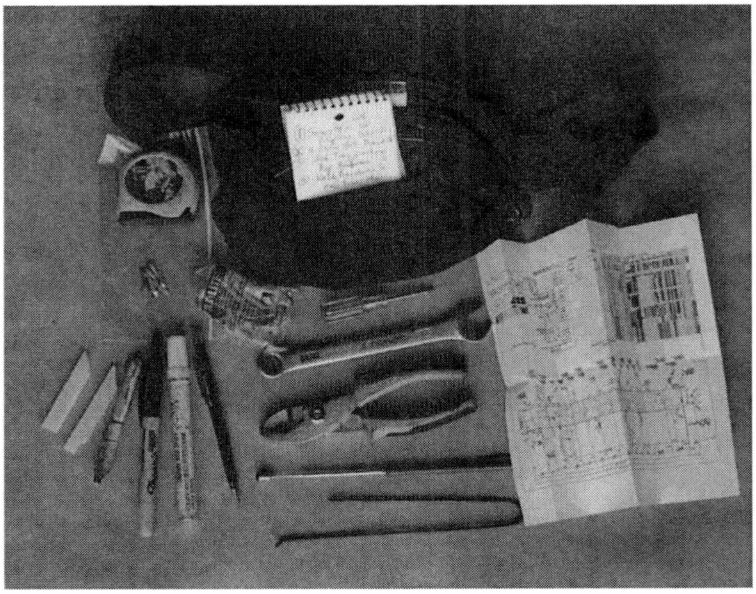

14

Other Equipment

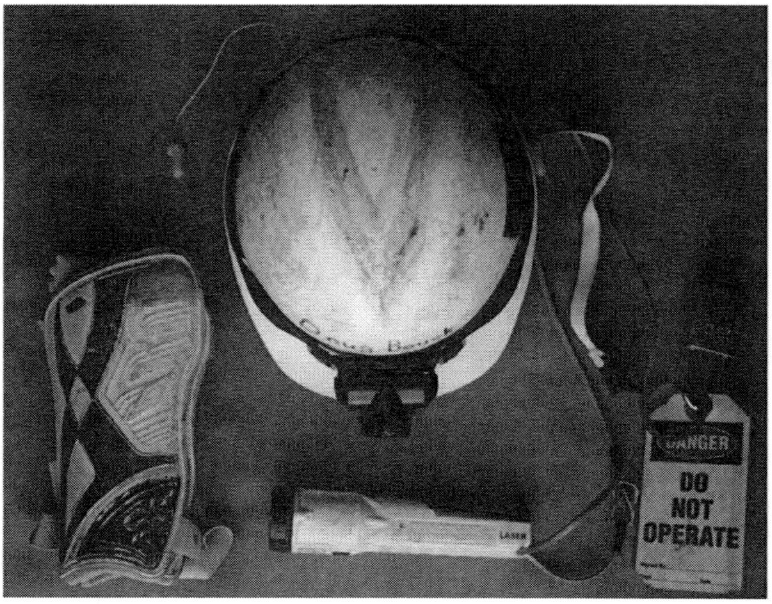

15

Inspection Equipment

- Special
  - Goggles, Dust Mask or Face Mask
  - Weir/Downcomer Dimension Template
  - 6 Inch Steel Rule
  - Dimension & Tolerance Table
  - Detailed Tray/Internals Drawings
  - Punch List and/or Drawings
  - Level Device: Bubble, Water or Laser
    - Device Accuracy Test
  - Distributor Test Equipment
  - Water Tightness Test Equipment

16

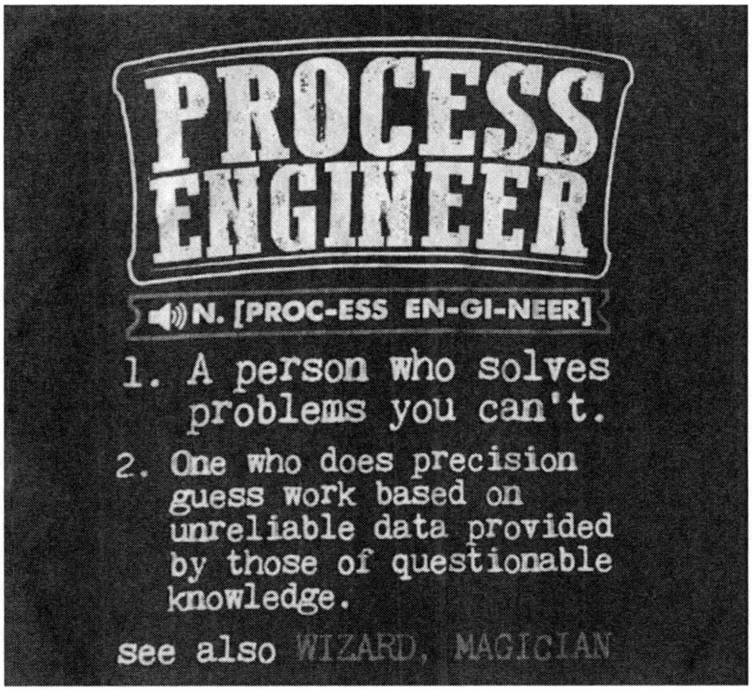

17

Keys to Process Inspection

- Understand How the Process Works
- Understand How the Parts Fit Together
- Understand How the Parts Work Together
- Know Why the Specs are Important
- Know How Deviations Affect the Process
- Be Able to Explain Operating and Economic Effects of Deviations
- Consider the Next Fix Chance May be 5 Years
  - Know When to Hold'em and When to Fold'em

18

Basic Preliminary Tray Inspection

- Number Trays for Reference
- Note Cleaning Issues – Take Photos
  - Does It Need Cleaning Before Full Prelim.?
  - Loose Deposits or Hard Scale?
  - Tray Decks/Downcomers and On/Under Valves
  - Any Special Cleaning Techniques Required?
- Note & Mark Missing/Stuck Valves
- Note & Mark Loose or Missing Hardware
- Note Cracked/Damaged/Corroded Trays
- Note Condition of Internal Manway Hardware
- Note Condition of Feed Pipes & Vortex Breakers
- Prepare Recommendations & Punch List

19

TAR Tray Inspection 1

- Objectives:
  - Return Equip. to Efficient Operation
  - Remember the Next Fix Opportunity May Be Years
- Cleanliness
  - Loose Scale/Deposits on Downcomers & Walls
    - Any Loose Scale Can Move to Downstream Equipment
  - Deposits on Decks
    - Plugging or Restriction of Valves, Holes or Caps
    - Bubble/Tunnel Caps Annular Space Must Be Clear
  - Accumulation in Downcomers, Sealpans & Drawpans

20

TAR Tray Inspection 2

- Missing/Worn/Stuck/Plugged Valves
  - One (1) Percent Acceptable Exception
  - Moving Valves Must Move Full Travel
    - Remove Deposits on Top, Legs or Around Holes of Valves
  - Check for Worn Valves & Tray Decks
  - Replacement Valves: Legs Bent for Retention
  - Replacement Valves with Cages: Crimping of Leg Tabs

21

Valve Cage Installation Problem

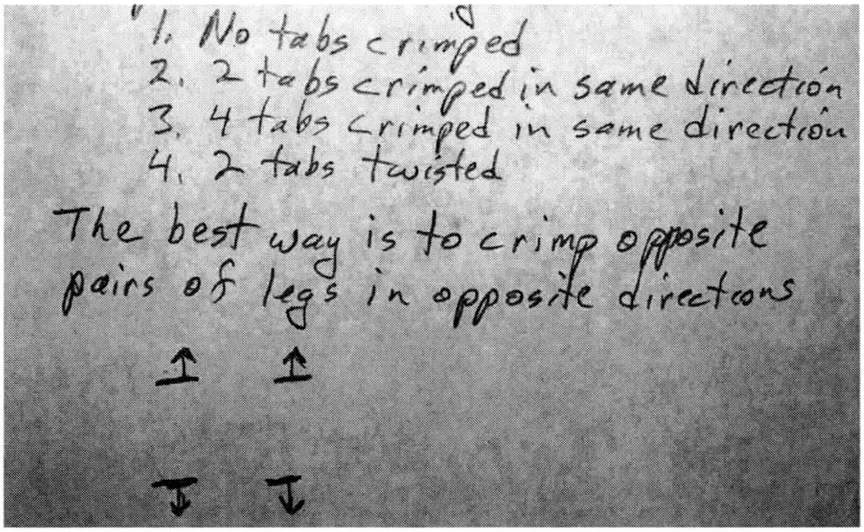

22

TAR Tray Inspection 3

- **Most Frequent Problem:**
  - Frozen/Damaged Internal Manway Hardware
    - Result of Hammer Release and/or Corrosion
    - Frozen Hardware Guarantees Poor Re-Closure
    - Poor Closure Causes Vibration Cracking of Trays
    - Poor Closure Causes Less Security Against Upset
    - Poor Closure May Result In Operating Deficiency
  - **Hardware Problems Must be Corrected Before Approval for Internal Closure**

23

TAR Tray Inspection 4

- Nutter MW Hardware Problems:
  - Subtle Correct Orientation
  - Rectangular Slots and Bolts
  - Boilermaker Hammer Release Damage
  - Same Poor Closure Issues
- Blind Fastening
  - Unusual
  - Difficult
  - Two Different Types of Hardware
  - Two Different Techniques
  - Boilermakers May Need Instruction

24

Nutter Internal Manway Hardware Orientation

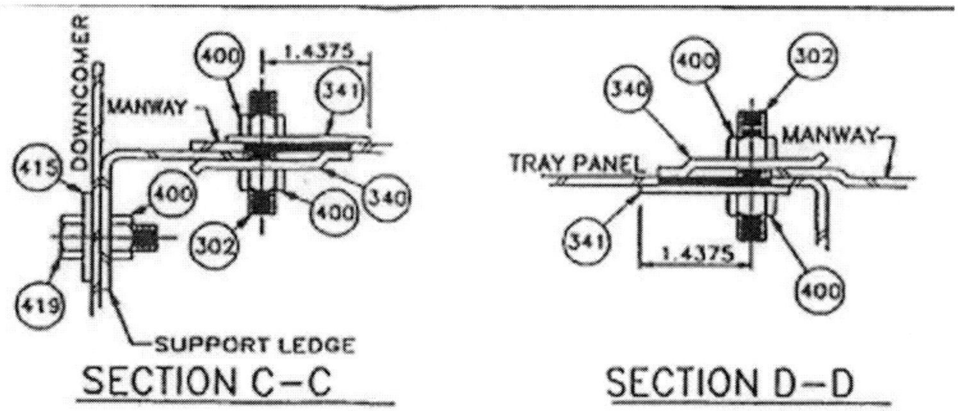

25

Blind Manway Bolting

26

Blind Manway Bolting

27

TAR Tray Inspection 5

- Other MW Hardware Problems
  - Light Wt. Football Washer Bends Causing Poor 2nd Closure
  - 3/16 in. Thick Heavy Duty Washers (Footballs) Recommended

28

Holes in Downcomers

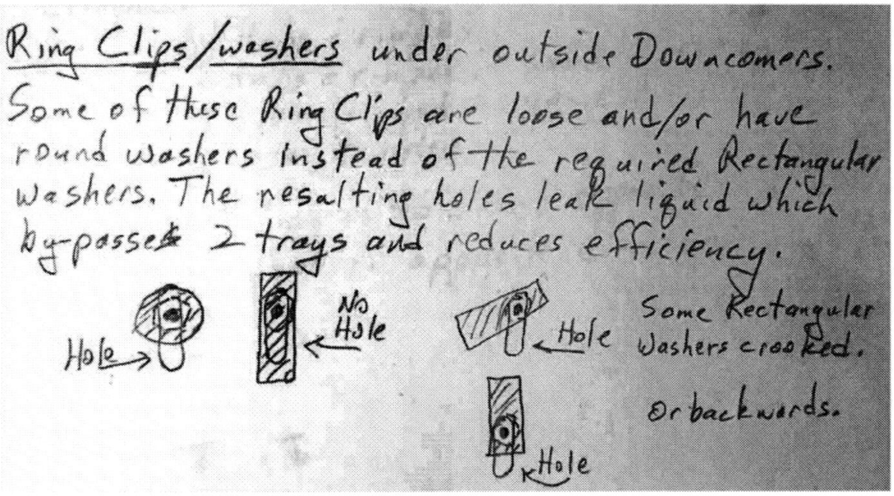

29

Internal Downcomer Loose Scale

30

Downcomer Scale

31

Dirty Round Valve Tray

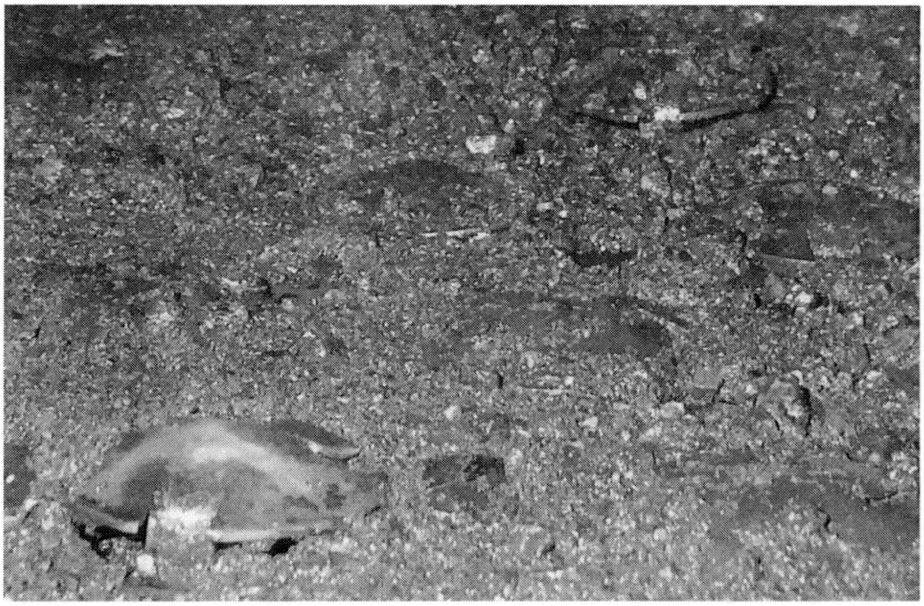

32

1- 1.5 Inch Deposits

33

Corroded and Plugged Bubble Caps

34

Displaced Weir

35

Bent/Loose Downcomer Clip

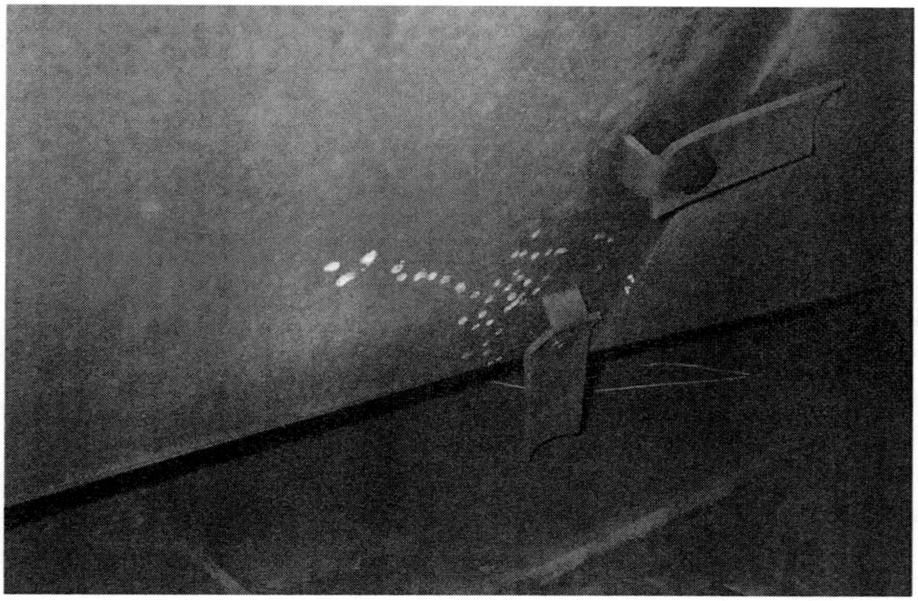

36

Bent Tray Downcomer & Clip

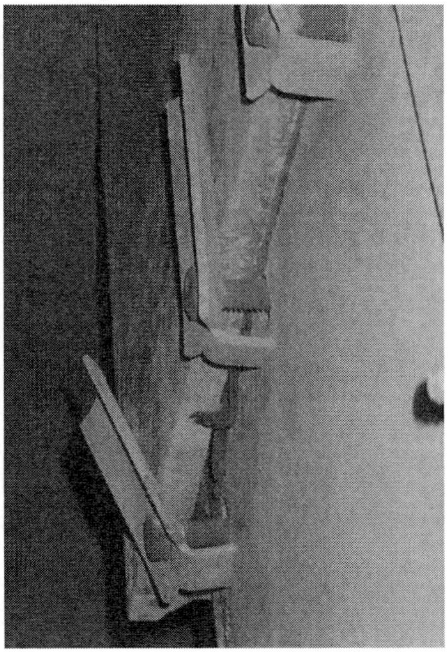

37

Ring Clip Askew

38

Displaced Valves

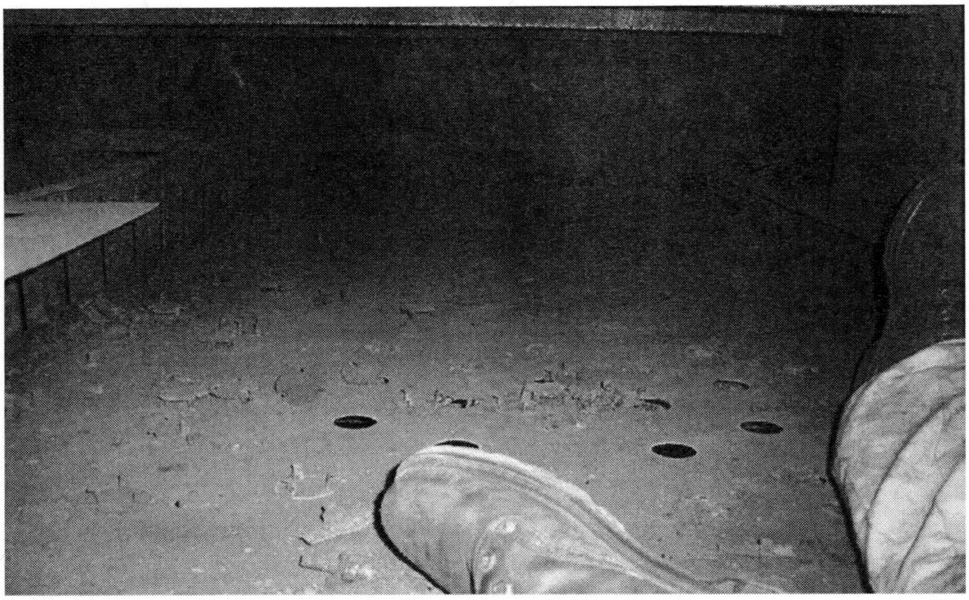

39

Displaced Valves

- Check the Legs Condition
  - Valve Legs Worn?
  - Improper Installation?
    - Legs not Spread Properly?
- Check the Hole Condition
  - Worn By Erosion-Corrosion?
- Blown Loose by Process Problem?
- Deck Thinning from Corrosion?

40

Missing Bubble Cap

41

Missing Swept Back Weir Piece

42

Missing Deck Bolts

43

Tray Damage Analysis

- Trays Bent or Cracked
  - Fatigued by Age and/or Weight
  - Overtightening
  - Level Loss or Explosion
- Trays Pushed Up
  - High Level + Percolating Vapor
  - Water Explosion Below
- Trays Pushed Down
  - High Level + Rapid Level Loss
  - Water Explosion Above

44

Boss says:

DOUGBERT, THE CEO SAYS THAT THE PROBLEM DIB COLUMN IS OPEN FOR INSPECTION AND YOU ARE THE MAN!

Dougbert says:

WHY ME? THAT TOWER IS OVER 200 FEET TALL!

Boss says:

HE SAYS YOU ARE THE ONLY ONE WHO KNOWS HOW THE INSIDES ARE SUPPOSED TO WORK.

45

Bent Tray Panel at Support Ring

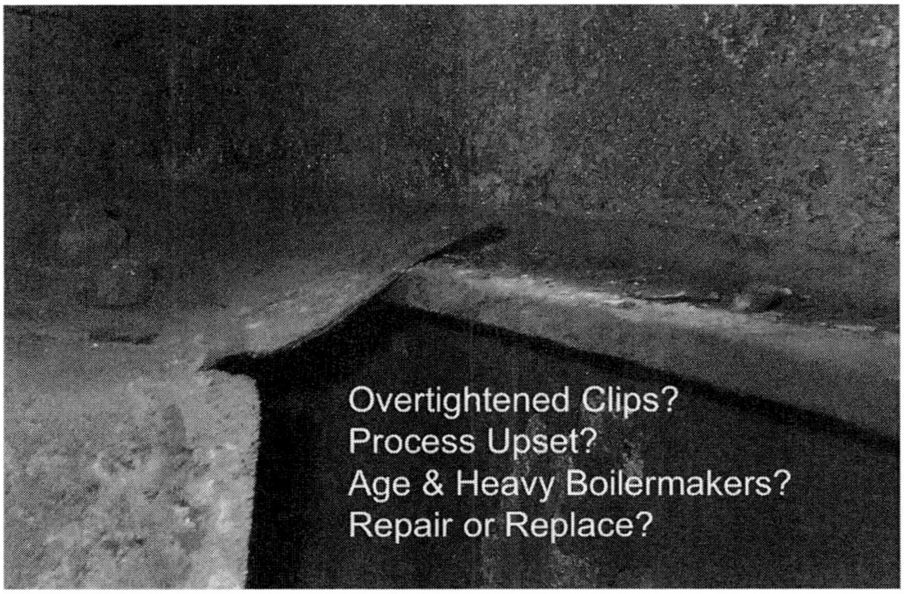

46

Results of Loss of Level Control with Vapor Pushing Liquid Up Through the Trays

47

Uplifted Downcomer Panel

48

Lift Damage

49

Lift Damage

50

Lift Damage

51

Push Down Damage

52

Push Down Damage

53

Brittle Failure

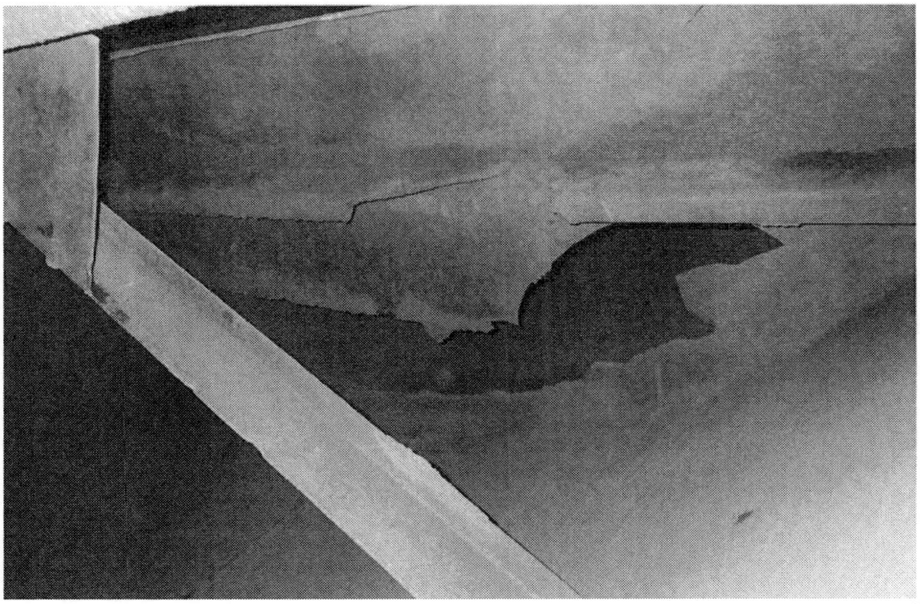

54

Process Revamp Review

- Pre-TAR Revamp Process Review
  - What is Different
  - Objective of Change
  - Impact of Change on the Process
  - Will the Changes Meet the Objectives
  - Are the Changes Consistent and Valid Engineering?
- Pre-TAR Mechanical Review
  - Mechanical Design meet Process Design?
  - Integration with Surrounding Systems: High Capacity Trays + Reboiler + Condenser?

55

Revamp Inspection

- Installed Mechanical & Process Check:
  - Installed per Drawings
  - DC Installation Blocks (Record DC Escapes & Any DC Changes)
  - Alignment of Tray Sections (Centering & Panel to Panel) Out of Round?
  - Correct Hardware Installation
  - Hardware Orientation & Proper Tightening
  - Punch List & Final Process Check

56

Revamp Field Inspection Examples

- Parts Verification Check: Example Valve Lift
- Small Parts Check and Security: Examples Twr. Closure Hold-up & Missing Nozzles
- Field Process Check: Example Missing Downcomers
- Field Dimension Check: Example Stripper Dbl. Flood
- Field Dimension Check: Support Ring Thickness
- Inspection Timing: Example Redesigned Feed Pipe
- Field Layout: Example Chimney Tray Water Draw

57

Example: Chimney Tray Water Draw

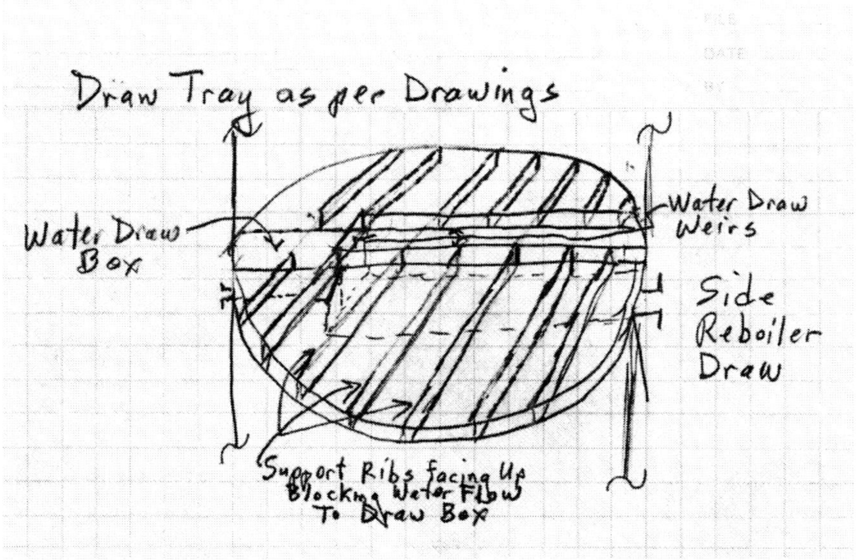

Bouck, Doug, Distillation Revamp Pitfalls to Avoid, Chemical Engineering Progress, p. 32-38, Feb. 2014

58

Special Inspections 1

- Feed and Return Pipe Distributors
  - Holes/Slots/Nozzles
    - Clear?
    - Corroded?
  - Flange Bolts Tight (Wrench Check)
  - Distributor Support Bolting Tight (Dbl. Nutting Use)
- False Downcomers & Distributor Impact Panels
  - Double Check for Loose Hardware
  - Need for extra security from vibration?
  - Vibration initiated cracking?

59

Distributor Hole Corrosion

60

Special Inspections 2

- Material Balance or Draw Problems?
- Flow or Level Control Problems?
- Suspected Leaks?
- Chimney Trays & Draw Boxes
  - Corrosion & Deposits in Draw and/or Piping
  - Thickness Checks in Suspect Corrosion Areas
  - Potential Leak Areas at Welds or Gasketing
  - Potential Leak Areas at Bottom (Under-deposit Corrosion)
  - Loose or Bent Hats, Chimneys or Decks

61

Chimney Draw Tray

62

Draw Box Corroded Through at Bottom

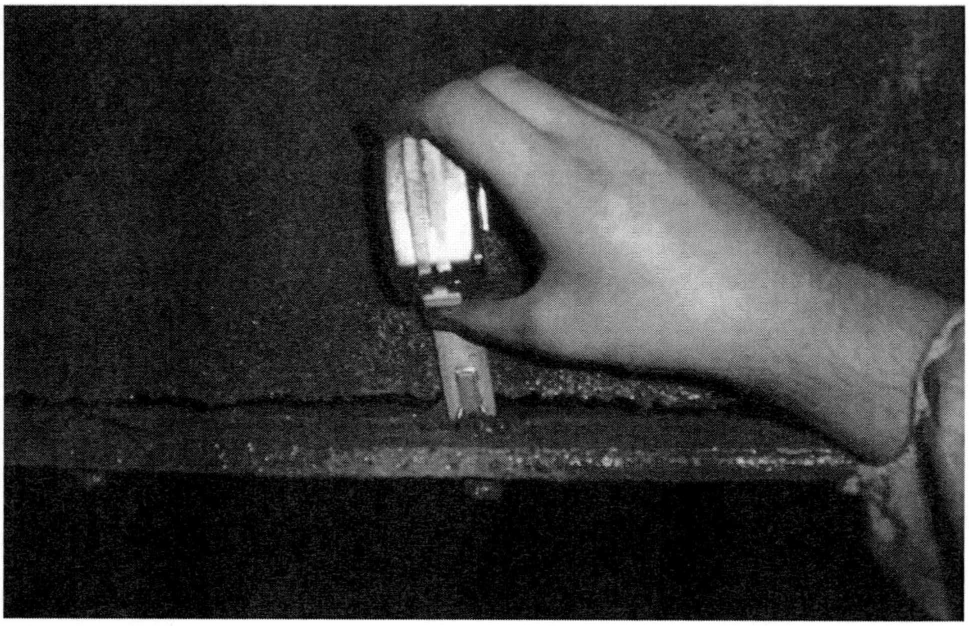

63

Draw Box Corroded Through at Welds

64

Special Inspections 3

- New or Revamped Chimney Trays
  - Gasketed or Welded
  - Leak Testing
    - Water Quality (Chlorides for Certain 300 Series Stainless and/or Solids in Firewater)
    - Quantity - Multiples Greater Than Leak Rate
    - Pressure - Pump Head Requirements (Water)
    - Test Depth - Approx. Operating Level
- Distributor Testing
  - On Site Spray Distributor Example
  - Off Site Packing Distributor Vendor Testing

65

Special Inspections 4

- Final Inspection Before Closure
  - All Internal Repairs & Cleaning Complete
  - Internal Manway Hardware Fixed or Replaced
  - Account For All Internal Manways – Directional?
  - Tools & Misc. Picked Up

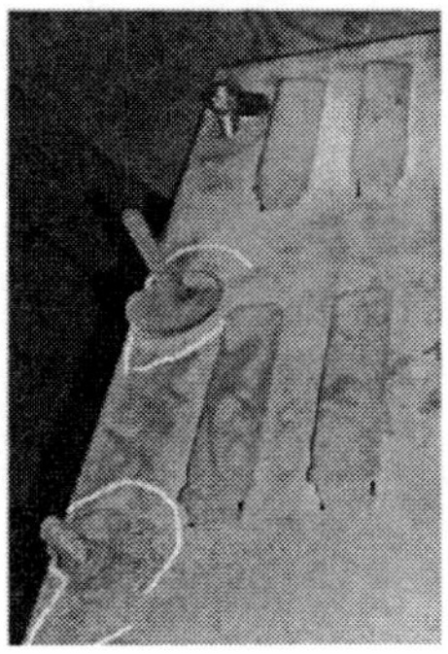

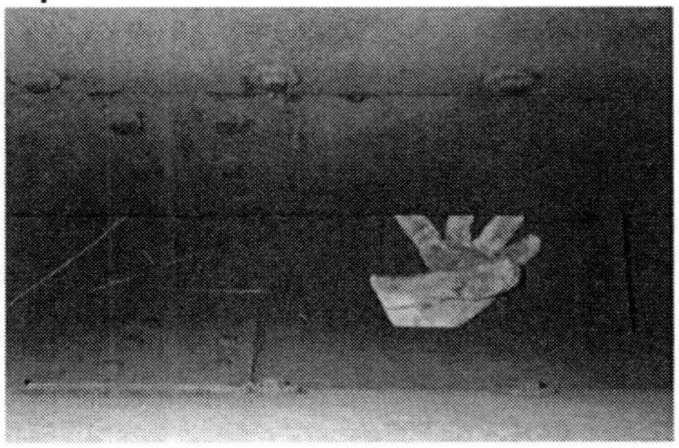

66

New Tower Inspection 1

- Tolerances
- Level Checks
  - Trays – Dependent on ring levelness
  - Weirs (If Adjustable)
- Downcomer Width – Top & Bottom
- Downcomer Escape
- Weir Heights (If Adjustable)
- Alignment/Centering of Tray Sections
- Valve Rise & Valve/Hole Counts

67

New Tower Inspection 2

- Hardware
  - Per Drawing
  - Correct Metallurgy
  - Bolting Snug Tight (~15 ft-lbs for 3/8 inch)
  - NOT Overtight
    - Bend Tray
    - Bend Clamps
    - Bend Washers
    - Gall Bolts

68

Pyrophoric Fire Prevention

- Pyrophoric Deposits
  - Iron Sulfide Can Be Most Anywhere Sulfur Is
  - Conversion Treatments Not Perfect
  - Prefer One-Step Processes Like ZymeFlow
  - Preventive Measures + Monitoring
  - FeS + O2, Dry and 104+ Degrees F. = Exothermic
  - Examples: Amine, Main Fractionators
- Carbon, Coke & Other Organics
- Hot Work
  - Read Chemical Engineering July 2007 Article on "Causes and Prevention of Packing Fires"

69

Packed System Inspection with NO Packing Removal

- No Specific Operating Problems
- Pre-TAR Gamma Scan
- Distributor Piping – Clean & Bolting Tight
- Pre-Distributor - Clean & Functional
- Distributor Pan, Troughs or Spray
  - Clean
  - Holes or Nozzles Clear
- Visual Inspection Top of Packing
- Visual Inspection Bottom of Packing

70

Structured Packing Installation

Feed Inlet Pipe
Predistrubutor
Distributor
Optional Holddown
Top Layer of Packing
90 Deg. Oriented Next Layer

Alternating Orientation of Layers

Bottom Layer of Packing
Packing Support
Liquid Collector

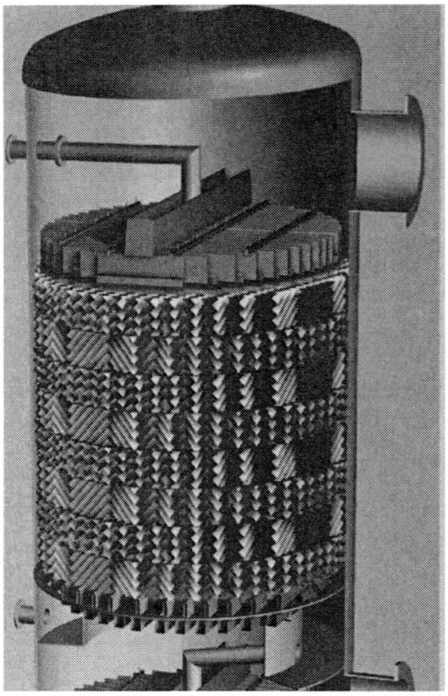

Sulzer Chemtech Brochure

71

Packed System Inspection With Packing Removal

- Before Re-Installation of Packing Check:
  - Draw Integrity – Check for Corrosion & Cleanliness
  - Support Pan/Grid – Clean/Unplugged
  - Check Support Orientation
  - Check Connections and Support Bolting Tightness
- Structured Packing Installation:
  - Verify Packing Size, Type & Metallurgy
  - Starting Orientation is Important
  - 90 Degree Alternate Layers (45 or 60 Degree Rotation for Grid)

72

Protect the Packing

- Walk on Planks NOT the Packing
- WHY for Structured Packing??
  - Structured Packing is Very Thin
  - Boots Smash the Top Crimps
  - Smashed Crimps Reduce Capacity
  - Smashed Crimps Reduce Efficiency
  - Hidden Damage is Untraceable
- Make Sure the Contractor is Complying

73

Protect the Packing

Stand on Planks
not the Packing

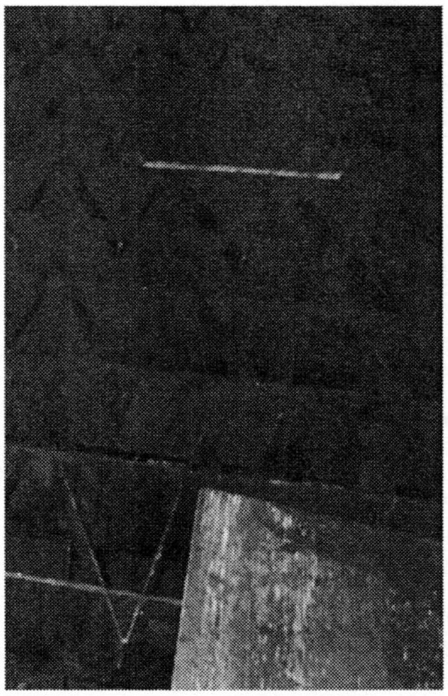

74

Protect the Hands/Arms

Koch-Glitsch Brochure

75

Installation

- Personal Safety
  - Gloves for Sharp Edges
- Orientation to Support
- Orientation Layer to Layer
- Shoehorns for Multi-piece Installation
- Distribute Excess Space Between Pieces
- Wall Wipers of Various Designs

76

Structured Packing Layers

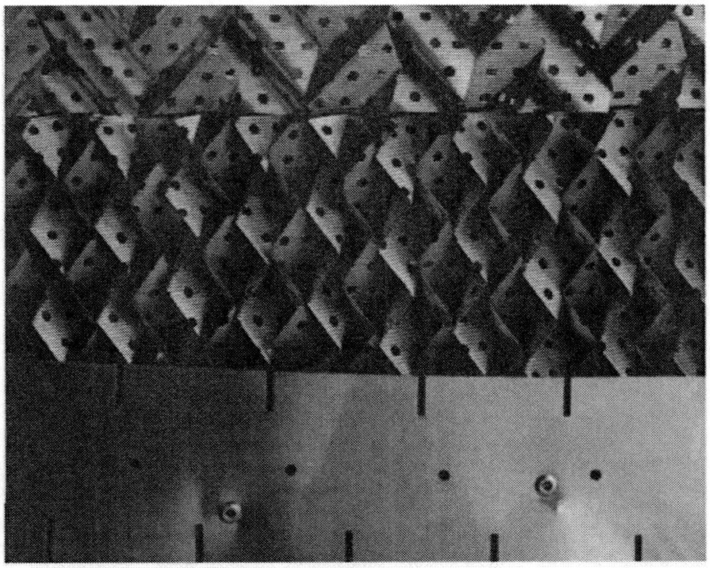

FLEXIPAC® Structured Packing with perforated and textured surface shown with attached wall wiper band

77

Other Wall Wipers

Sulzer Mellapac Structured Packing

78

Holddown Grid – Miss-installed

79

Distributor Inspections

- Holddown – Bolting Tight
- Level Distributors & Pre-Dist. 1/8 in. High to Low
- Installation Hold Point for Level Check/ Adjustments
- All Distributor and Predistributor Bolting Tight
- Distributor Clean
- Predistributor Clean
- Feed Piping Re-installed & Bolted Tight

80

Spray Distributors

- Check all Spray Nozzles and Headers
  - Remove Nozzles (Have Spares on Hand)
  - View Nozzles Internals – Clean or Replace
  - Save Nozzles in Safe Place
  - Check and Clean Headers as Necessary
  - Wrench Check all Nozzles for Tightness
  - Wrench Check all Flange Bolting
  - Double Nut and Check all Header Support Bolting

81

Detached Spray Headers

82

Upset Structured Packing (No Holddown Grid)

83

Redistributor Between Packed Beds

84

Random Packing Installation

Feed or Transfer Pipe
Predistributor
Distributor
Retaining Grid

Random Packing

Support Plate
Support Ring

Vapor Inlet

Liquid Level Collector/Sump

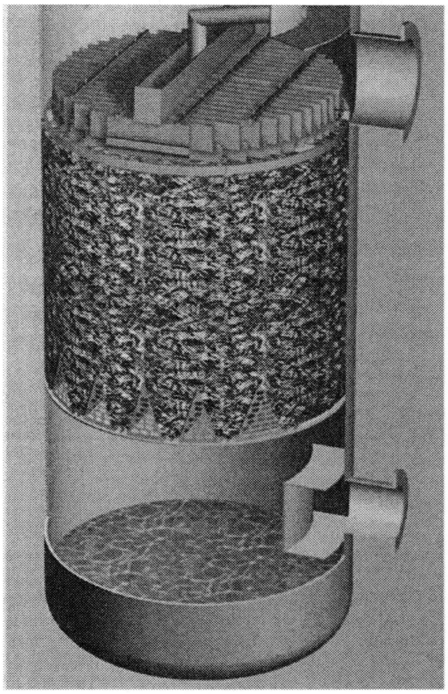

Sulzer Chemtech Brochure

85

Packing - Random

- Removal for Replacement
  - Remove Piping, Distributors & Retaining Grid
    - Safe Storage of All Parts
    - Record Orientations for Re-Installation
    - Clean All Parts
  - Remove Rings by Vacuum
  - Clean Any Collector or Draw System
  - Clean Packing Support
  - Check Support Integrity & Bolting

86

Packing - Random

- Installation
  - Verify Packing Size, Type and Metallurgy
  - Dump New Packing
  - Rake Top
  - **Do Not Overfill!**
  - If Necessary to Enter, Stand on Planks not Packing
  - Re-install Retainer Grid
  - Level Distributors & Pre-Dist. 1/8 in. High to Low
    - Installation Hold Point for Level Check/Adjustments
  - Install Feed Piping
  - Check Tightness of All Bolting

87

Protect the Packing

- Walk on Planks NOT the Packing
- WHY for Random Packing??
  - Random Packing is Thin
  - Boots Crush or Compress the Rings
  - Hidden Damage is Untraceable
  - Crushed Rings Reduce Capacity
  - Crushed Rings Reduce Efficiency
- Make Sure the Contractor is Complying
- Stay Totally Off, If Possible

88

The Process Engineer
(A Bit of Philosophy)

- Little Authority
- Much Responsibility
- Bridge Operations, Maintenance & Management
- Optimization Economics - Beyond Politics
- Relationships Create Influence
- Influence Creates Accepted Recommendations
- Recommendations Create Improvements
- Improvements Create More Profits
- More Profits Create Stable Company and Jobs

89

Take Ownership of the Future. The Role of Process Engineers in Inspection **Is** Important!

2 Peter 1:15

And I will make every effort to see that after my departure you will always be able to remember these things.

YOU ARE the FUTURE!

90

481654 Distillation Tower Inspection for Turnarounds and Revamps: Tutorial and Presentation - Part II

Abstract:

This presentation of "Distillation Tower Inspection for Turnarounds and Revamps" covers the unique safety issues for Distillation Tower inspections plus typical work plan responsibilities and inspection equipment. Both preliminary and follow-up inspection of trays and packed systems are covered. Emphasis is placed on what to look for in typical and unique inspection situations with both trayed and packed towers. Examples will be demonstrated and/or discussed including the unique perspective that the Process Engineer brings to the inspection process especially in replacement and revamp situations.

AUTHOR INDEX